Families of Bentham and Hooker Classification

MONOCOTYLEDONS

V. Darani M.Sc., M.Phil., SET

Chapter 1

Bentham and Hooker's classification

Chapter 1: Bentham and Hooker Classification (1862-1883)

Division: Phanerogams or Seed Plants

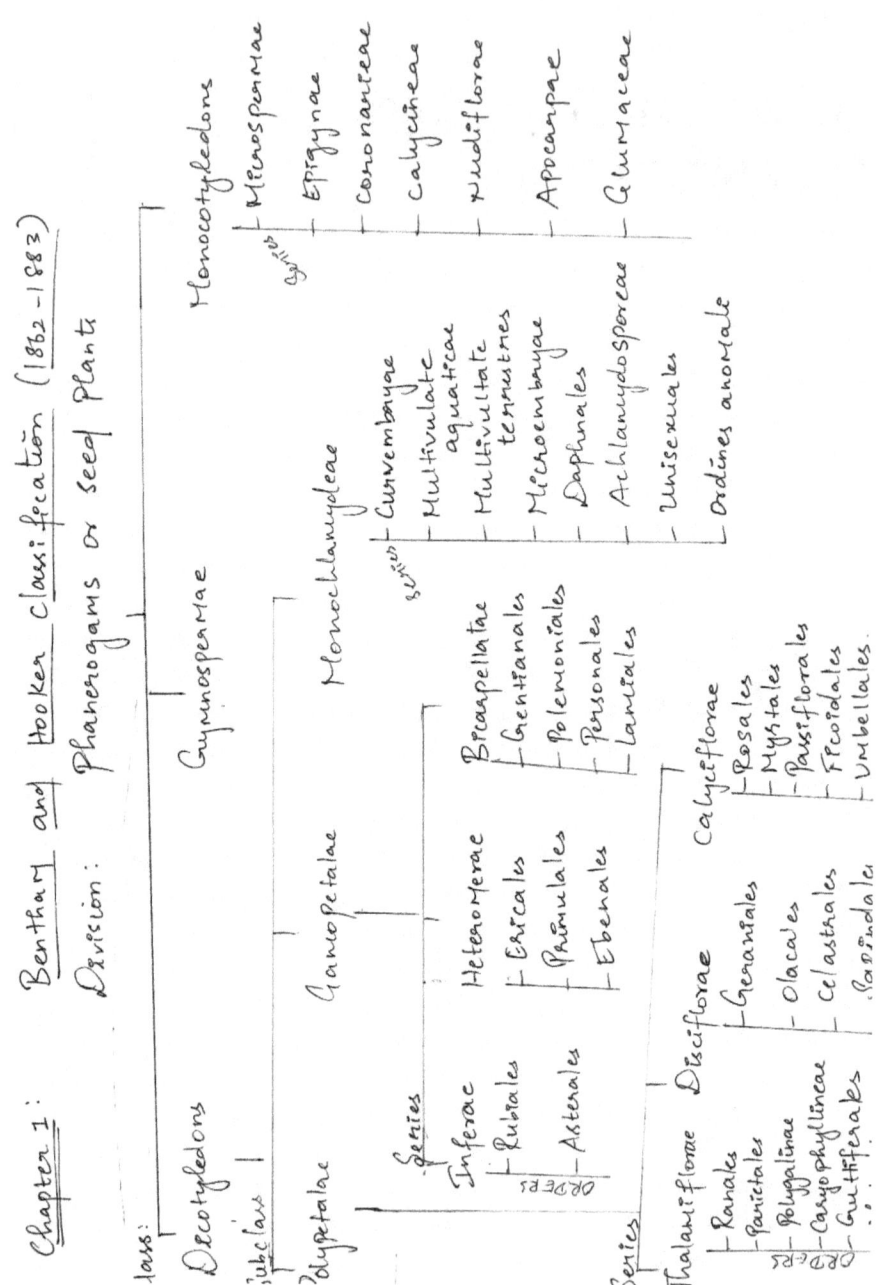

- **Class: Dicotyledons**
  - **Subclass:**
    - **Polypetalae**
      - *Series:* **Thalamiflorae**
        - Orders: Ranales, Parietales, Polygalinae, Caryophyllineae, Guttiferales
      - *Series:* **Disciflorae**
        - Geraniales, Olacales, Celastrales, Sapindales
      - *Series:* **Calyciflorae**
        - Rosales, Myrtales, Passiflorales, Ficoidales, Umbellales
    - **Gamopetalae**
      - *Series:* **Inferae**
        - Orders: Rubiales, Asterales
      - **Heteromerae**
        - Ericales, Primulales, Ebenales
      - **Bicarpellatae**
        - Gentianales, Polemoniales, Personales, Lamiales
    - **Monochlamydeae**
      - *series:* Curvembryae, Multiovulatae aquaticae, Multiovulatae terrestres, Micrembryae, Daphnales, Achlamydosporeae, Unisexuales, Ordines anomale
  - **Gymnospermae**
- **Class: Monocotyledons**
  - Microspermae, Epigynae, Coronarieae, Calycineae, Nudiflorae, Apocarpae, Glumaceae
  (series)

Families of Monocotyledons

Families listed in Bentham and Hooker's Classification

Subclass: Polypetalae
Series: Thalamiflorae

Order: Ranales
Families: 1) Ranunculaceae 2) Dilleniaceae 3) Calycanthaceae 4) Magnoliaceae 5) Annonaceae 6) Menispermaceae 7) Berberidaceae 8) Nymphaeaceae

Order: Parietales
Families: 1) Sarraceniaceae 2) Papaveraceae 3) Cruciferae 4) Capparidaceae 5) Resedaceae 6) Cistineae 7) Violaceae 8) Canellaceae 9) Bixineae

Order: Polygalineae
Families: 1) Pittosporeae 2) Tremandreae 3) Polygaleae 4) Vochysiaceae

Order: Caryophyllineae
Families: 1) Frankeniaceae 2) Caryophyllaceae 3) Portulacaceae 4) Tamariscineae

Order: Guttiferales
Families: 1) Elatineae 2) Hypericineae 3) Guttiferae 4) Ternstroemiaceae 5) Dipterocarpaceae 6) Chlaenaceae

Order: Malvales
Family: 1) Malvaceae 2) Sterculiaceae 3) Tiliaceae
Series: Disciflorae
Order: Geraniales
Families: 1) Lineae 2) Humiriaceae 3) Malphigiaceae
4) Zygophyllaceae 5) Geraniaceae 6) Rutaceae
7) Simarubeae 8) Ochnaceae 9) Burseraceae
10) Meliaceae 11) Chailletiaceae

Order: Olacales
Families: 1) Olacineae 2) Ilicineae 3) Cyrilleae
Order: Celastrales
Families: 1) Celastrineae 2) Stackhousieae 3) Rhamneae
4) Ampelideae
Order: Sapindales
Families: 1) Sapindaceae 2) Sabiaceae 3) Anacardiaceae
4) Coriarieae 5) Moringeae

Series: Calyciflorae
Order: Rosales
Families: 1) Connaraceae 2) Leguminosae
3) Rosaceae 4) Saxifrageae 5) Crassulaceae
6) Droseraceae 7) Hamamelideae 8) Bruniaceae
9) Halorageae
Order: Myrtales
Families: 1) Rhizophoraceae 2) Combretaceae

Families of Monocotyledons

3) Myrtaceae 4) Melastomaceae 5) Lythrarieae
6) Onagrarieae

Order: Passiflorales
Families: 1) Samydaceae 2) Loaseae 3) Turneraceae
4) Passifloreae 5) Cucurbitaceae 6) Begoniaceae
7) Datisceae

Order: Ficoidales
Families: 1) Cactaceae 2) Ficoideae

Order: Umbellales
Families: 1) Umbelliferae 2) Araliceae 3) Cornaceae

Subclass: Gamopetalae

Series: Inferae
Order: Rubiales
Families: 1) Caprifoliaceae 2) Rubiaceae

Order: Asterales
Families: 1) Valerianeae 2) Dipsaceae 3) Calycereae
4) Compositae.

Series: Heteromerae
Order: Ericales
Family: 1) Ericaceae 2) Vaccinieae 3) Monotropeae
4) Epacrideae 5) Diapensiaceae 6) Lennoaceae

Order: Primulales
Family: 1) Plumbagineae 2) Primulaceae
3) Myrsineae

Order: Ebenales
Family: 1) Sapotaceae 2) Ebenaceae 3) Styraceae
Series: Bicarpellatae
Order: Gentianales
Family: 1) Oleaceae 2) Salvadoraceae 3) Apocynaceae
4) Asclepiadaceae 5) Loganiaceae 6) Gentianaceae
Order: Polemoniales
Family: 1) Polemoniaceae 2) Hydrophyllaceae
3) Boragineae 4) Convolvulaceae 5) Solanaceae
Order: Personales
Family: 1) Scrophularineae 2) Orobanchaceae
3) Lentibularieae 4) Columelliaceae 5) Gesneraceae
6) Bignoniaceae 7) Pedalineae 8) Acanthaceae
Order: Lamiales
Family: 1) Myoporineae 2) Selagineae
3) Verbenaceae 4) Labiatae 5) Plantagineae

Subclass: Monochlamydeae (Incompletae)
Series: Curvembryae
Family: 1) Nyctagineae 2) Illecebraceae
3) Amarantaceae 4) Chenopodiaceae 5) Phytolaccaceae
6) Batideae 7) Polygonaceae
Series: Multiovulatae Aquaticae
Family: 1) Podostemaceae
Series: Multiovulatae Terrestres

7 | Families of Monocotyledons

Family: 1) Nepenthaceae 2) Cytinaceae 3) Aristolochieae

Series: Micrembryae
Family: 1) Piperaceae 2) Chloranthaceae 3) Myristiceae 4) Moniaceae

Series: Daphnales
Family: 1) Laurineae 2) Proteaceae 3) Thymelaceae 4) Penaeaceae 5) Eleagnaceae

Series: Achlamydosporae
Family: 1) Loranthaceae 2) Santalaceae 3) Balanophoreae

Series: Unisexuales
Family: 1) Euphorbiaceae 2) Balanopseae 3) Urticaceae 4) Plantanaceae 5) Leitnerieae 6) Juglandeae 7) Myricaceae 8) Casuarineae 9) Cupuliferae

Series: Ordines anomali
Family: 1) Salicaceae 2) Lacistemaceae 3) Empetraceae 4) Ceratophylleae

Monocotyledons

Series: Microspermae
Family: 1) Hydrocharideae 2) Burmanniaceae 3) Orchideae

Series: Epigynae
Family: 1) Scitamineae 2) Bromeliaceae 3) Haemodoraceae 4) Irideae 5) Amaryllideae 6) Taccaceae 7) Dioscoreaceae.

Series: Coronariae.
Family: 1) Roxburghiaceae 2) Liliaceae 3) Pontederiaceae 4) Philydraceae 5) Xyrideae 6) Mayacaceae 7) Commelinaceae 8) Rapateaceae

Series: Calycineae
Family: 1) Flagellariaceae 2) Juncaceae 3) Palmae

Series: Nudiflorae
Family: 1) Pandaneae 2) Cyclanthaceae 3) Typhaceae 4) Aroideae 5) Lemnaceae

Series: Apocarpae
Family: 1) Triurideae 2) Alismaceae 3) Najadaceae

Series: Glumaceae
Family: 1) Eriocauleae 2) Centrolepideae 3) Restiaceae 4) Cyperaceae 5) Gramineae.

Series: Microspermae

Families:

Hydrocharideae

Burmanniaceae

Orchideae

Hydrocharideae

Phanerogams
Monocotyledons
Microspermae
Hydrocharideae
Hydrocharitaceae

General characters:

→ Small aquatic herbs, either free floating or submerged.

→ Stem - reduced, sending out numerous spreading fibrous roots and stolons.

→ At the end of stolon new leaf-rosettes arise

→ *Elodea* bears an elongated, slender branched stem bearing whorls of narrow toothed leaves. In *Vallisneria* the shortened stem bears a tuft of roots and a crowded cluster of long, linear leaves. *Hydrochari* possesses rosettes of rounded kidney-shaped leaves.

→ Generally the leaves are basal and often crowded or cauline, when cauline they are alternate, opposite or whorled, usually sessile.

→ Inflorescence is always lateral. Usually there is one inflorescence at each axil. The young inflorescence remains enclosed within a spathe formed by a two free or somewhat united bracts which very often persist upto the maturation of the fruit.

→ Flowers are unisexual but rarely hermaphrodite. Plants may be monoecious or dioecious. In *Halophil* both male and female flowers are found within the same spathe. The male inflorescence consist of one to many flowers. The female and bisexual inflorescence are generally one flowered.; genera actinomorphic, rarely zygomorphic, trimerous and epigynous.

→ Perianth is white and easily distinguished into outer calyx and inner corolla. Petals remain inserted on the united base of the sepals. Calyx is tough and protective in function whereas corolla is fugacious and delicate. Sepals are quite small, and the petals are broad and membranous. The female flowers generally possess staminodes, but the male flowers do not have any trace of pistil. The flowers of *Halophila* do not have petals and the perianth is represented by 3 sepals.

→ Androecium consists of 4 whorls, the innermost whorl comprises of only staminodes.

Families of Monocotyledons

while in the third whorl there are half-anther and two whorls having 3 stamens in each whorl.

→ Gynoecium consists of 2-15 carpels, syncarpous, ovary inferior, unilocular with Parietal placentation. Placentas are often produced in the centre. Generally numerous ovules.

→ Fruit - berry like, undehiscent, leathery or fleshy, submerged and opening irregularly.

→ Seeds - Many, non-endospermic and containing a large embryo.

→ Pollination - Hydrophilous.

Genera
→ Apalanthe
→ Appertiella
→ Blyxa
→ Egeria
→ Elodea
→ Enhalus
→ Halophila
→ Hydrilla
→ Hydrocharis
→ Lagarosiphon
→ Limnobium
→ Maidenia
→ Nechamandra
→ Ottelia
→ Stratiotes
→ Thalassia
→ Vallisneria

Burmanniaceae

Classification (Bentham and Hooker)

- Phanerogams
- Monocotyledons
- Microspermae
- Burmanniaceae

General characters

→ Herbs, annual or perennial, mycotrophic, semi-mycotrophic or autotrophic

→ Often Rhizomatous or tuberous.

→ Leaves - alternate, simple, entire; with cauline leaves; autotrophic species with rosette leaves

→ Inflorescence terminal, many flowered cymes or Racemes. or in some cases flowers solitary.

→ Flowers - bisexual, Perianth of 1 or 2 whorls and each whorl with 3 tepals, corolline, tubular or campanulate. Perianth tube often 3 angled or 3 winged; tepals sometimes appendaged; appendage terminal, elongated, slender. Stamens 3 or 6; if 3 then subsessile in perianth throat, if 6 then pendent in perianth tube; connectives large, often appendiculate. Inferior ovary, 1-loculed, Parietal placentation or 3-loculed with axile Placentation; ovules numerous, anatropous; style filiform, shortly cylindric or conic; stigmas 3, sometimes connate.

→ Fruit capsule, often fleshy, with either persistent Perianth tube and style or only persistent basal ring of Perianth, dehiscence irregular or by transverse ventral slits.

→ Seeds small, numerous; endosperm present

Genera included under Burmanniaceae

- → Apteria
- → Burmannia
- → Campylosiphon
- → Cymbocarpa
- → Dictyostega
- → Gymnosiphon
- → Hexapterella
- → Marthella
- → Miersella
- → Afrothismia
- → Haplothismia
- → Oxygyne
- → Thismia
- → Tiputinia

Orchidaceae

Classification (Bentham and Hooker)

Phanerogams
Monocotyledons
Microspermae
Orchideae

General Characters

→ Vary in habit. Always herbaceous with sympodial stems. Terrestrial herbs, many are epiphytes and several others are saprophytes. Many are xerophytes and they store water in thickened leaves, aerial roots, swollen internodes and Pseudobulbs. Epiphytic orchids develop long aerial roots for the absorption of water and are called velamen, a tissue consisting of several layers of short tracheids. The other roots perform the function of anchorage.

→ Possess short rhizome. Each annual shoot bends up into the leafy branch of the current year. In many species, the internode of the stem become thickened and is known as Pseudobulb.

→ Leaves — simple, alternate, rarely opposite or whorled, which may be fleshy. In many, they are reduced to scales which may be membranous or succulent. Usually linear, often distichous and sometimes radical. Very often leaf sheath develops which surround the stem. The foliage leaf may be entirely absent, when the shoot is reduced to scaly bulb. Margin entire.

→ Inflorescence – Racemose type, spike, raceme, Panicle. Sometimes flowers solitary (cypripedium). In species of <u>Bulbophyllum</u>, the flowers remain almost sunk in a fleshy axis. Panicle in <u>Oncidium</u>. Generally inflorescence is short lived but rarely becomes perennial producing fresh flowers round the year.

→ Flower – hermaphrodite, sometimes unisexual, medianly zygomorphic, bracteate, sessile or pedicellate and epigynous. Flowers often very colourful and attractive, but rarely colourless or brown or inconspicuous.

→ Perianth – 6 perianth leaves arranged in 2 whorls of 3 each. Outer whorl is calyx like and inner whorl forms the corolla. Perianth leaves may be free or somewhat united. Segments are generally imbricate. Generally the whole perianth becomes twisted through 180° with the result of a torsion produced by the ovary, and then the labellum in the anterior abaxial position. Labellum is the enlargement and modification of one of the segments of inner whorl.

→ Androecium – 2 trimerous whorls of stamens, which are never all present. Generally there are 1 or 2 stamens, which unite together and to an extension of the gynoecium forming a column. Anthers – bicelled, introrse and dehisce by longitudinal slits. Pollinia present.

→ Gynoecium – tricarpellary, syncarpous; ovary inferior, unilocular, very rarely trilocular. When unilocular, placentation is parietal. Stigma – modified. *Cypripedium* – all the 3 lobes of the stigma are functional, but more often the anterior lobe of the stigma becomes sterile and develops into a pocket known as rostellum, the remaining 2 stigmatic lobes are functional. A part of rostellum develops into a viscid disc known as viscidium.

→ Fruit – capsule, opening by 3 to 6 longitudinal slits.

→ Seeds – numerous, very minute, blown away by wind.

→ Pollination – entomophilous.

Genera included under Orchidaceae

- Aa
- Abdominea
- Acacallis
- Acampe
- Acanthephippium
- Aceras
- Aceratorchis
- Acianthus
- Acineta
- Ackermania
- Acoridium
- Acostaea
- Acriopsis
- Acrochaene
- Acrolophia
- Acrorchis
- Ada
- Adenochilus
- Adenoncos
- Adrorhizon
- Aerangis
- Aeranthes
- Aerides
- Aganisia
- Aglossorrhyncha
- Agrostophyllum
- Alamania
- Altensteinia
- Amblyanthe
- Ambrella
- Amerorchis
- Amesiella
- Amitostigma
- Amparoa
- Anacamptis
- Ancistrochilus
- Ancistrorhynchus
- Androcorys
- Angraecopsis
- Angraecum
- Anguloa
- Anoectochilus
- Ansellia
- Anteriorchis
- Anthogonium
- Anthosiphon
- Antillanorchis
- Aorchis
- Aphyllorch
- Aplectrum
- Aporostylis
- Aporum
- Apostasia
- Appendicula
- Aracamunia
- Arachnis
- Archineottia
- Arethusa
- Armodorum
- Arnottia
- Arpophyllum
- Arthrochilus
- Artorima
- Arundina
- Ascidieria
- Ascocentrum
- Ascochilopsis
- Ascochilus
- Ascoglossum
- Ascolabium
- Aspasia

Families of Monocotyledons

- Aspidogyne
- Aulosepalum
- Auxopus
- Baptistonia
- Barbosella
- Barbrodria
- Barkeria
- Barlia
- Bartholina
- Basigyne
- Basiphyllaea
- Baskervilla
- Batemannia
- Beclardia
- Beloglottis
- Benthamia
- Benzingia
- Biermannia
- Bifrenaria
- Binotia
- Bipinnula
- Bletia
- Bletilla
- Bogoria
- Bolbidium
- Bollea
- Bolusiella
- Bonatea
- Bonniera
- Brachionidium
- Brachtia
- Brachycorythis
- Brachypeza
- Brachystele
- Bracisepalum
- Braemia
- Brassavola
- Brassia
- Briegeria
- Bromheadia
- Broughtonia
- Brownleea
- Buchtienia
- Bulbophyllum
- Bulleyia
- Burnettia
- Burnsbaloghia
- Cadetia
- Caladenia
- Calanthe
- Caleana
- Callostylis
- Calochilus
- Calopogon
- Caluera
- Calymmanthe
- Calypso
- Calyptrochilus
- Campanulorch
- Campylocentrum
- Capanemia
- Cardiochilus
- Catasetum
- Cattleya
- Cattleyopsis
- Caucaea
- Caularthron
- Centroglossa
- Centrostigma
- Cephalanthera
- Cephalantheropsis
- Ceratandra
- Ceratocentron
- Ceratochilus
- Ceratostylis

- Chaemaeangis
- Chamaeanthus
- Chamaegastrodia
- Chamelophyton
- Chamorchis
- Changnienia
- Chaseella
- Chaubardia
- Chaubardiella
- Chauliodon
- Cheiradenia
- Cheirostylis
- Chelonistele
- Chiloglottis
- Chilopogon
- Chiloschista
- Chitonanthera
- Chitonochilus
- Chloraea
- Chondradenia
- Chondrorhyncha
- Chroniochilus
- Chrysocycnis
- Chrysoglossum
- Chusua
- Chysis
- Chytroglossa
- Cirrhaea
- Cleisocentron
- Cirrhopetalum
- Cischweinfia
- Cladenia
- Cleisomeria
- Cleisostoma
- Cleistes
- Clematepistephium
- Clowesia
- Coccineorchis
- Cochleanthes
- Cochleola
- Cocleorchis
- Codonorchis
- Codonosiphon
- Coelia
- Coeliopsis
- Coeloglossum
- Coelogyne
- Coilochilus
- Collabium
- Comparettia
- Comperia
- Conchidium
- Condylago
- Constantia
- Corallorrhiza
- Cordiglottis
- Coryanthes
- Corybas
- Corycium
- Corymborkis
- Corysanthes
- Cottonia
- Cotyloabium
- Cranichis
- Cremastra
- Cribbia
- Crossoglossa
- Cryptarrhena
- Cryptocentrum
- Cryptochilus
- Cryptopus

Families of Monocotyledons

- Cryptopylos
- Cryptostylis
- Cuitlauzina
- Cyanaeorchis
- Cybebus
- Cyclopogon
- Cycnoches
- Cylindrolobus
- Cymbidiella
- Cymbidium
- Cymboglossum
- Cynorkis
- Cyphochilus
- Cypholoron
- Cyrtidiorchis
- Cyrtopodium
- Cystorchis
- Cyrtosia
- Cystostylis
- Cystorchis
- Dactylorhiza
- Dactylorhynchus
- Dactylostalix
- Degranvillea
- Deiregyne
- Dendrobium
- Dendrochilum
- Dendrophylax
- Diadenium
- Diaphananthe
- Diceratostele
- Dicerostylis
- Dichaea
- Dichromanthus
- Dickasonia
- Dictyophyllaria
- Didiciea
- Didymoplexiella
- Didymoplexis
- Diglyphosa
- Dignathe
- Dilochia
- Dilochiopsis
- Dilomilis
- Dimerandra
- Dimorphorchis
- Dinema
- Dinklageella
- Diothonea
- Diphylax
- Diplandrorchis
- Diplocaulobium
- Diplocentrum
- Diplolabellum
- Diplomeris
- Diploprora
- Dipodium
- Dipteranthus
- Dipterostele
- Disa
- Discyphus
- Dispensis
- Distylodon
- Dithyridanthus
- Diuris
- Dockrillia
- Dodsonia
- Dolichocentrum

→ Domingoa
→ Doritis
→ Dossinia
→ Dracula
→ Drakaea
→ Dresslerella
→ Dressleria
→ Dryadella
→ Dryadorchis
→ Drymoanthus
→ Drymoda
→ Duckeella
→ Dunstervillea
→ Dyakia
→ Earina
→ Eggelingia
→ Eleorchis
→ Elleanthus
→ Eloyella
→ Eltroplectris
→ Elythranthera
→ Embreea
→ Encyclia

→ Entomophobia
→ Eparmatostigma
→ Ephippianthus
→ Epiblastus
→ Epiblema
→ Epicranthes
→ Epidanthus
→ Epidendrum
→ Epigeneium
→ Epilyna
→ Epipactis
→ Epipogium
→ Epistephium
→ Eria
→ Eriaxis
→ Eriochilus
→ Eriodes
→ Eriopexis
→ Eriopsis
→ Erycina
→ Erythrodes
→ Erythrorchis
→ Esmeralda

→ Euanthe
→ Eucosia
→ Eulophia
→ Eulophiella
→ Euphlebium
→ Eurycentrum
→ Eurychone
→ Eurystyles
→ Evotella
→ Fernandezia
→ Ferruminaria
→ Fimbriella
→ Flickingeria
→ Frondaria
→ Fuertesiella
→ Funkiella
→ Galeandra
→ Galearis
→ Galeola
→ Galeottia
→ Galeottiella
→ Garaya
→ Gastrochilus
→ Gastrodia

23 | Families of Monocotyledons

- Gastrorchis
- Gavilea
- Geesinkorchis
- Gennaria
- Genoplesium
- Genyorchis
- Geoblasta
- Geodorum
- Glomera
- Glossodia
- Glossorhyncha
- Gomesa
- Gomphichis
- Gonatostylis
- Gongora
- Goniochilus
- Goodyera
- Govenia
- Gracielanthus
- Grammangis
- Grammatophyllum
- Graphorkis
- Grastidium
- Greenwoodia

- Gorodya
- Grosourdya
- Gularia
- Gunnarella
- Gunnarorchis
- Gymnadenia
- Gymnadeniopsis
- Gymnochilus
- Gymnoglottis
- Habenaria
- Hagsatera
- Hammarbya
- Hancockia
- Hapalochilus
- Hapalorchis
- Harrisella
- Hederorkis
- Helcia
- Helleriella
- Helonoma
- Hemipilia
- Herminium
- Herpetophytum
- Herpysma

- Herschelianthe
- Hetaeria
- Heterozeuxine
- Hexalectris
- Hexisea
- Himantoglossum
- Thitonella
- Hippeophyllum
- Hirtzia
- Hispaniella
- Hoehneella
- Hofmeisterili
- Holcoglossum
- Holopogon
- Holothrix
- Homalopetalum
- Horichia
- Hormidium
- Horvatia
- Houlletia
- Huntleya
- Huttonaea
- Hybochilus
- Hygrochilus

- Hylophila
- Hymenorchis
- Imerinaea
- Inobulbon
- Ione
- Ionopsis
- Ipsea
- Isabelia
- Ischnocentrum
- Ischnogyne
- Isochilus
- Isotria
- Jacquiniella
- Jejosephia
- Jumellea
- Kalimpongia
- Kefersteinia
- Kegeliella
- Kerigomnia
- Kinetochilus
- Kingidium
- Kionophyton
- Koellensteinia
- Konantzia
- Kreodanthus
- Kryptostoma
- Kuhlhasseltia
- Lacana
- Laelia
- Laeliocattleya
- Laeliopsis
- Lanium
- Lankesterella
- Leaoa
- Lecanorchis
- Lemboglossum
- Lemurella
- Lemurorchis
- Leochilus
- Lepanthes
- Lepanthopsis
- Lepidogyne
- Leporella
- Leptotes
- Lesliea
- Leucohyle
- Ligeophila
- Limodorum
- Liparis
- Listera
- Listrostachys
- Lockhartia
- Loefgrenianthus
- Ludisia
- Lueddemannia
- Luisia
- Lycaste
- Lycomormium
- Lyperanthus
- Lysoglossa
- Macodes
- Macradenia
- Macroclinium
- Macropodanthus
- Malaxis
- Malleola
- Manniella
- Margellianthus
- Masdevallia
- Mastigion
- Maxillaria
- Mediocalcar

- Megalorchis
- Megalotus
- Megastylis
- Meiracyllium
- Mendoncella
- Mesadenella
- Mesadenus
- Mesoglossum
- Mesospinidium
- Mexicoa
- Microcoelia
- Micropera
- Microtatorchis
- Microthelys
- Microtis
- Miltonia
- Miltoniopsis
- Mischobulbum
- Mobilabium
- Moerenhoutia
- Monadenia
- Monanthos
- Monomeria
- Monophyllorchis
- Monosepalum
- Mormodes
- Mormolyca
- Mycaranthes
- Myoxanthus
- Myrmechis
- Myrmecophila
- Myrosmodes
- Mystacidium
- Nabaluia
- Nageliella
- Nematoceros
- Neobathiea
- Neobenthamia
- Neobolusia
- Neoclemensia
- Neocogniauxia
- Neodryas
- Neoescobaria
- Neofinetia
- Neogardneria
- Neogyna
- Neomoorea
- Neotinea
- Neottia
- Neottianthe
- Neowilliamsia
- Nephelaphyl
- Nephrangis
- Nervilia
- Neuwiedia
- Nidema
- Nigritella
- Nothodorit
- Nothostele
- Notylia
- Oberonia
- Octarrhena
- Octomeria
- Odontochilu
- Odontoglos.
- Odontorrhynch!
- Oeceoclades
- Oeonia
- Oeoniella
- Oerstedella
- Olgasis
- Oligophyton
- Oliveriana

- Omoea
- Oncidium
- Ophidion
- Ophrys
- Orchipedum
- Orchis
- Oreorchis
- Orestias
- Orleanesia
- Ornithocephalus
- Ornithochilus
- Ornithophora
- Orthoceras
- Osmoglossum
- Ossiculum
- Osyricera
- Otochilus
- Otoglossum
- Otostylis
- Palatia
- Pachites
- Pachyphyllum
- Pachyplectron
- Pachystele
- Pachystoma
- Palmorchis
- Palumbina
- Panisea
- Pantlingia
- Paphinia
- Papilionanthe
- Papillilabium
- Papperitzia
- Papuaea
- Paradisanthus
- Paraphalaenopsis
- Parapteroceras
- Pecteilis
- Pedilochilus
- Pedilonum
- Pelatantheria
- Pelexia
- Pennilabium
- Peristeranthus
- Peristeria
- Peristylus
- Pescatoria
- Phaius
- Phalaenopsis
- Pholidota
- Phragmorchis
- Phreatia
- Phymatidium
- Physoceras
- Physogyne
- Pilophyllum
- Pinelia
- Piperia
- Pityphyllum
- Platanthera
- Platycoryne
- Platyglottis
- Platylepis
- Platyrhiza
- Platystele
- Platythelys
- Plectorrhiza
- Plectrelminthus
- Plectrophora
- Pleione
- Pleurothallis
- Pleurothallopsis

- Plocoglottis
- Poaephyllum
- Podangis
- Podochilus
- Pogonia
- Pogoniopsis
- Polycycnis
- Polyotidium
- Polyradicion
- Polystachya
- Pomatocalpa
- Ponera
- Ponerorchis
- Ponthieva
- Porolabium
- Porpax
- Porphyrodesme
- Porphyroglottis
- Porphyrostachys
- Porroglossum
- Porrorhachis
- Prasophyllum
- Prescottia
- Pristiglottis
- Promenaea
- Protoceras
- Pseudacoridium
- Pseuderia
- Pseudocentrum
- Pseudocranichis
- Pseudoeurystyles
- Pseudogoodyera
- Pseudolaelia
- Pseudorchis
- Pseudovanilla
- Psilochilus
- Psychilis
- Psychopsiella
- Psychopsis
- Psygmorchis
- Pteraichis
- Pteroceras
- Pteroglossa
- Pteroglossapis
- Pterostemma
- Pterostylis
- Pterygodium
- Pygmaeorchis
- Pyrorchis
- Quekettia
- Quisqueya
- Rangaeris
- Rauhiella
- Raycadenco
- Reichenbachthu
- Renanthera
- Renantherella
- Restrepia
- Restrepiella
- Restrepropsis
- Rhaesteria
- Rhamphorhync
- Rhinerrhiza
- Rhizanthella
- Rhynchogyn
- Rhyncholael
- Rhynchophrea
- Rhynchoste
- Rhynchostyl
- Rhytionante
- Rodleyella
- Rimacola

- Risleya
- Robiquetia
- Rodriguezia
- Rodrigueziella
- Rodrigueziopsis
- Roeperocharis
- Rossioglossum
- Rudolfiella
- Rusbyella
- Saccoglossum
- Saccolabiopsis
- Saccolabium
- Sacoila
- Salpistele
- Sanderella
- Sarcanthopsis
- Sarcochilus
- Sarcoglottis
- Sarcoglyphis
- Sarcophyton
- Sarcostoma
- Satyridium
- Satyrium
- Saundersia
- Sauroglossum
- Scaphosepalum
- Scaphyglottis
- Scelochiloides
- Scelochilus
- Schiedeella
- Schistotylus
- Schizochilus
- Schizodium
- Schlimmia
- Schoenorchis
- Schomburgkia
- Schwartzkopffia
- Scuticaria
- Sedirea
- Seidenfadenia
- Sepalosiphon
- Serapias
- Sertifera
- Sieve Kingia
- Sigmatostalix
- Silvorchis
- Sirhookera
- Sirhookera
- Skeptrostachys
- Smithorchis
- Smithsonia
- Smitinandia
- Sobennikoffia
- Sobralia
- Solenangis
- Solenidiopsis
- Solenidium
- Solenocentrum
- Sophronitella
- Sophronitis
- Soterosanthus
- Spathoglottis
- Sphyrarhynchos
- Sphyrastylis
- Spiculaea
- Spiranthes
- Stalkya
- Stanhopea
- Staurochilus
- Steles
- Stellilabium
- Stenia
- Stenocoryne
- Stenoglottis
- Stenoptera
- Stenorrhynchos

→ Stephanothelys
→ Stereochilus
→ Stereosandra
→ Stevensella
→ Stictophyllum
→ Stigmatosema
→ Stolzia
→ Suarezia
→ Summerhayesia
→ Sunipia
→ Sutrina
→ Svenkoeltzia
→ Symphyglossum
→ Synanthes
→ Synarmosepalum
→ Systeloglossum
→ Taeniophyllum
→ Taeniorrhiza
→ Tainia
→ Tangtsinia
→ Tapeinoglossum
→ Taprobaena
→ Telipogon

→ Tetragamestus
→ Tetramicra
→ Teuscheria
→ Thaia
→ Thecopus
→ Thecostele
→ Thelasis
→ Thelychiton
→ Thelymitra
→ Thelyschista
→ Thrixspermum
→ Thulinia
→ Thunia
→ Thysanoglossa
→ Ticoglossum
→ Tipularia
→ Tolumnia
→ Townsonia
→ Trachyrhizum
→ Traunsteinera
→ Trevoria
→ Trias
→ Triceratorhynchus
→ Trichocentrum

→ Trichoceros
→ Trichoglottis
→ Trichotosia
→ Tridactyle
→ Trigonidium
→ Triphora
→ Trisetella
→ Trizeuxis
→ Tropidia
→ Trudelia
→ Tsaiorchis
→ Tuberolabium
→ Tubilabium
→ Tulotis
→ Tylostigma
→ Uleiorchis
→ Uncifera
→ Urostachys
→ Vanda
→ Vandopsis
→ Vanilla
→ Vargasiella
→ Vasqueziella

- Ventricularia
- Vesicisepalum
- Vexillabium
- Vrydagzynea
- Wallnoeferia
- Warmingia
- Warrea
- Warreella
- Warreopsis
- Warscaea
- Wullschlaegelia
- Xenikophyton
- Xerorchis
- Xiphosium
- Xylobium
- Yoania
- Ypsilopus
- Zeuxine
- Zootrophion
- Zygopetalum
- Zygosepalum
- Zygostates

Series: Epignae

Families:

Scitmineae

Bromellaceae

Haemodoraceae

Irideae

Amaryllideae

Taccaceae

Dioscoreaceae

Musaceae (Scitamineae)

Classification (Bentham and Hooker)

- Phanerogams
- Monocotyledons
- Epigynae
- Scitamineae
- Musaceae

General Characters

→ Mostly Perennial herbs of big dimensions. They are tree like in appearance and hence are termed as gigantic herbs. Most of them possess underground branched rhizome and a Pseudoaerial stem.

→ Adventitious Root

→ Stem - underground rhizome or root stock. Pseudoerect aerial stem reach upto 10-15 feet height is actually formed by the long, stiff and overlapping leaf sheaths. The erect aerial shoots die after they bear fructifications. The apparent aerial stem is termed as 'shaft'.

→ Leaves- simple, sometimes very large upto 6 ft or more, glabrous with a strong midrib and lateral parallel veins. The leaves possess large and overlapping sheaths; petiolate. Petiole is long and thick. The leaves are rolled up in early stages but later on they spread in large laminas.

→ Inflorescence - Flowers are borne in terminal spikes or panicles which are covered by spathes and the floral clusters often found in the axils of large and showy bracts. In Musa, the large red bracts are arranged in 3 spiral lines round the peduncle, each bract covers several unisexual flowers which arise without any bract from the axil of the spike. At the opening of the flowers, the bracts roll back and ultimately fall off.

→ Flowers - Sessile, hermaphrodite, sometimes unisexual; zygomorphic, epigynous. When unisexual, the plants are monoecious. The male and female flowers are arranged on the same inflorescence, the male flowers are found within the upper bracts while the female flowers are within the lower bracts towards the base of the inflorescence. The flowers are complete when hermaphrodite and incomplete when unisexual.

→ Perianth leaves are found to be arranged in 2 whorls of 3 each.; gamophyllous or polyphyllous. In *Musa*, the 3 outer and 2 inner anterior Perianth leaves are united into a tube like structure. The inner perianth leaf is free. Tepals are coloured.

→ Androecium – 6 stamens, free, usually arranged in 2 whorls; 5 stamens are perfect while the 6th one is either absent or staminode. *Musa* – 6th stamen is absent. Anthers – bicelled, basifixed, introrse.

→ Gynoecium – 3 carpels, syncarpous; ovary inferior, trilocular, axile placentation, many ovules in each locule; style filiform; 3 branched stigmas.

→ Fruit – an elongated fleshy, 3 chambered, indehiscent berry.

→ Seeds – Exalbuminous with a straight embryo.

Genera included under Musaceae

→ <u>Ensete</u>   → <u>Musella</u>

→ <u>Musa</u>

Bromeliaceae

Classification (Bentham and Hooker)

Phanerogams
Monocotyledons
Epigynae
Bromeliaceae

General Characters

→ Herbs or few arborescent, Perennial
→ Plants usually more or less succulent or non-succulent.
→ Basal aggregation of leaves; mostly acaulescent rosette plants, adapted to absorb the water which accumulates in the vase shaped leaf rosettes; or with terminal aggregation of leaves or without conspicuous aggregation of leaves (sometimes with leaves on elongated stems) or with terminal aggregation of leaves.
→ Self supporting or epiphytic (commonly) or rarely climbing.
→ Leaves alternate, spiral (mostly) or distichous;

usually leathery and fleshy or modified into spines; usually imbricate, sessile, sheathing. Leaf sheaths with free margins. Leaves simple; Lamina – entire; linear or lanceolate or ovate or subulate, parallel venation; exstipulate, margin entire or serrate.

→ Plants usually hermaphrodite. Floral nectaries present. Nectar secretion from gynoecium via septal nectaries.

→ Inflorescence – racemes, spikes or heads.
→ Flowers – bracteate; bracts distichous, often conspicuous and brightly coloured; regular or somewhat irregular; trimerous; cyclic; pentacyclic. Perigone tube present or absent.
→ Perianth with distinct calyx and corolla; calyx 3, 1 whorl, poly or gamosepalous, imbricate. Corolla 3, 1 whorl, appendiculate (with coronal structures and/or paired, basal nectary scales), or not appendiculate; poly or gamopetalous. Corolla lobes markedly longer than tube. Corolla imbricate or contorted; green or white or yellow or orange or red or blue.

→ Androecium consists of 6 stamens; diplostemonous, alterniperianthial. Anthers dorsifixed or basifixed, dehiscence via longitudinal slits; introrse.

→ Gynoecium - 3 carpelled; carpels isomerous with the perianth. Pistil 3 celled. Gynoecium syncarpous, syn.stylovarious; superior to inferior; ovary 3 locular. Odd carpel anterior. Styles 1, apical; stigmas 3; commissural; wet or dry type; papillate. Placentation axile; ovules 5-50 per locule., anatropous.

→ Fruit - fleshy or non-fleshy; dehiscent or indehiscent; capsule or berry. Capsule - septicidal or loculicidal.

→ Seeds endospermic with well differentiated embryo.

Genera included under Bromeliaceae
- → Abromeitiella
- → Acanthostachys
- → Aechmea
- → Alcantarea
- → Ananas
- → Androlepis
- → Araeococcus
- → Ayensua
- → Billbergia
- → Brewcaria
- → Brocchinia
- → Bromelia
- → Canistropsis
- → Canistrum
- → Catopsis
- → Chevaliera
- → Connellia
- → Cottendorfia
- → Cryptanthus
- → Deinacanthon
- → Deuterocohnia
- → Disteganthus
- → Dyckia
- → Edmundao
- → Encholirium
- → Fascicularia

- → Fernseea
- → Fosterella
- → Glomeropitcairnia
- → Greigia
- → Guzmania
- → Hechtia
- → Hohenbergia
- → Hohenbergiopsis
- → Lamprococcus
- → Lendmania
- → Lymania
- → Macrochondion
- → Mezobromelia
- → Navia
- → Neoglaziovia
- → Neoregelia
- → Nidularium
- → Ochagavia
- → Ortgiesia
- → Orthophytum
- → Pepinia
- → Pitcairnia
- → Platyaechmea
- → Podaechmea
- → Portea
- → Pseudaechmea
- → Pseudananas
- → Puya
- → Quesnelia
- → Racinaea
- → Ronnbergia
- → Steyerbromelia
- → Streptocalyx
- → Tillandsia
- → Ursulaea
- → Vriesea
- → Werhauia
- → Wittrockia

Haemodoraceae

Classification (Bentham and Hooker)

Phanerogams
Monocotyledons
Epigynae
Haemodoraceae

General characters:

→ Herbs with or without coloured juice
→ Perennial
→ Basal aggregation of leaves (radical); rhizomatous or bulbaceous or tuberous.
→ Leaves - alternate, distichous, leathery, sessile, sheathing. Leaf sheaths with free margins. Simple; Pulvinate; lamina entire; linear; Parallel venation; Margin entire
→ Plants hermaphrodite, Floral nectaries present. Nectar secretion from the gynoecium via septal nectaries.
→ Inflorescence or solitary; If inflorescence - cymes, Panicles, racemes.
→ Flowers - regular to very irregular, when irregular, zygomorphic, trimerous, cyclic,

tri or tetra cyclic. Perigone tube straight or curved or in some cases absent.

→ Perianth consists of tepals; 6 free or joined; 2 whorled or 1 whorled; isomerous; petaloid; similar in 2 whorls; green or white or cream or yellow or orange or violet or red or black.

→ Androecium — 3 or 6 - equal, 1 or 2 whorled; stamens 3 or 6, isomerous with the perianth, or diplostemonous. Anthers dorsifixed or basifixed versatile or not; dehisce via longitudinal slits; introrse.

→ Gynoecium — 3 carpelled; syncarpous or eu-syncarpous superior to inferior, styles 1 or 3, when 3 partially joined, attenuate from the ovary, apical, usually filiform, stigmas 1 or 3, capitate. Axile placentation ovules 1-50 per locule.

→ Fruit — non-fleshy, dehiscent or indehiscent, a capsule or nut

→ Seeds endospermic (oily)

<u>Genera</u>

→ Anigozanthos
→ Barberetta
→ Blancoa
→ Conostylis
→ Dilatris
→ Haemodorum
→ Lachnanthes
→ Macropidia
→ Xiphidium
→ Phlebocarya
→ Pyrrhorhiza
→ Schiekia
→ Tribonanthes
→ Wachendorfia

Irideae (Iridaceae)

Classification (Bentham and Hooker)

- Phanerogams
- Monocotyledons
- Epigynae
- Irideae

General Characters

→ Herbs or rarely shrubs.
→ Perennial; with a basal aggregation of leaves.
→ Leaves persistent or deciduous.
→ Leaves alternate, distichous, flat or terete, herbaceous or leathery, sessile or petiolate. Leaf sheaths with free or jointed margins.
→ Plants - hermaphrodite; Floral nectaries mostly present or rarely absent. Nectar secretion from the perianth or rarely from Gynoecium.
→ Flowers - solitary or aggregated in inflorescence - Panicles, cymes, spikes, umbels or corymbs. Spatheate (via one or 2 expanded, bladeless sheath)
→ Flowers bracteate, small to large, regular to very irregular; when irregular, zygomorphic, Trimerous, cyclic - tetracyclic. Long or short

- Perigone tube is present
- Perianth is represented by tepals – 6 in 2 whorls, isomerous, petaloid, commonly spotted rarely not spotted; either white, yellow, red, purple, violet or blue.
- Androecium – coherent – filaments often united into a basal tube, when united, monadelphous. 1 whorl. Stamens 2-3, Anthers separate or rarely cohering; basifixed; dehiscing via longitudinal slits; extrorse.
- Gynoecium – tricarpellary; partly petaloid or non-petaloid; syncarpous; synstylovarious; inferior or superior; ovary – unilocular or trilocular. Styles 1 or 3 or 6, Stylar canal present; placentation when unilocular is ~~parietal~~ or always axile; ovules 1-50 per locule.
- Fruit – non-fleshy, dehiscent, loculicidal capsule
- Seeds endospermic (oily)

<u>Genera included</u>
- Aninea
- Alophia
- Anomatheca
- Aristea
- Babiana
- Barnardiella
- Belamcanda
- Bobartia
- Calydorea
- Cardenanthus
- Chasmanthe
- Cipura

43 | Families of Monocotyledons

- → Cobana
- → Crocosmia
- → Crocus
- → Cypella
- → Devia
- → Dierama
- → Dietes
- → Diplarrhena
- → Duthiastrum
- → Eleutherine
- → Ennealophus
- → Ferraria
- → Fosteria
- → Freesia
- → Galaxia
- → Geissorhiza
- → Gelasine
- → Geosiris
- → Gladiolus
- → Gynandriris
- → Herbertia
- → Hermodactylus
- → Hesperantha
- → Hesperoxiphion
- → Hexaglottis
- → Homeria
- → Homoglossum
- → Iris
- → Isophysis
- → Ixia
- → Kelissa
- → Klattia
- → Lapeirousia
- → Lethia
- → Libertia
- → Mastigostyla
- → Pheome
- → Roggeveldia
- → Romulea
- → Savannosiphon
- → Schizostylis
- → Sessilanthera
- → Sisyrinchium
- → Solenomelus
- → Sparaxis
- → Sympa
- → Syringodea
- → Tapeina
- → Therelanthus
- → Tigridia
- → Trimezia
- → Tritonia
- → Tritoniopsis
- → Tucma
- → Watsonia
- → Wisenia
- → Zygotritonia

Amaryllidaceae

Classification (Bentham and Hooker)

    Phanerogams
    Monocotyledons
    Epigynae
    Amaryllideae

General characters

→ Plants are generally scapose, perennial herb with a tunicated bulb or rhizome.

→ Leaves are few in number and arise from the bulb (radical). More or less linear and have parallel venation.

→ Inflorescence – umbellate inflorescence. Rarely solitary. Borne at the top of a leafless scape. Flowers are subtended by an involucre of one or more membranous bracts.

→ Flowers – Pedicellate, bracteate, hermaphrodite, actinomorphic, trimerous, hypo or epigynous.

→ Perianth – 6 petaloid perianth leaves arranged in 2 whorls of 3 each. Very often, a corona is present and the perianth leaves are frequently united to form a tubular structure.

→ Androecium – 6 stamens. Stamens are found to

be inserted opposite the perianth leaves. In Gethyllis, there are 12-18 stamens. They are hypogynous and inserted on a tube at the base of the perianth leaves. Filaments are either free or connate at the base forming a corona. Anthers - bicelled, introrse, versatile or basifixed, dehisce by longitudinal slits.

→ Gynoecium - tricarpellary, syncarpous, superior or inferior. Axile placentation; ovary - trilocular. Many ovules per loculus; anatropous; style - slender with a capitate or trilobed stigma.

→ Fruit - capsule or may be fleshy and indehiscent

→ Seeds - small, numerous; endospermic

→ Pollination - entomophilous.

Genera included under Amaryllidaceae

- Amaryllis
- Ammocharis
- Apodolirion
- Bokkeveldia
- Boophone
- Bravoa
- Brunsvigia
- Caliphruria
- Calostemma
- Carpolyza
- Chlidanthus
- Choananthus
- Clivia
- Cooperia
- Crinum
- Cryptostephanus
- Cybistetes
- Cyrtanthus
- Eucharis
- Eucrosia
- Eustephia
- Galanthus
- Gemmaria
- Gethyllis
- Griffinia
- Habranthus
- Haemanthus
- Hannonia
- Hessea
- Hieronymiella
- Hippeastrum
- Hymenocallis
- Ismene
- Lapiedra
- Leptochiton
- Leucojum
- Lycoris
- Namaquanula
- Narcissus
- Nerine
- Pamianthe
- Pancratium
- Paramongaia
- Phaedranassa
- Phycella
- Placea
- Proiphys
- Pucara
- Pyrolirion
- Rauhia
- Rhodophia
- Scadoxus
- Sprekelia
- Stenomesson
- Sternbergia
- Strumaria
- Tapeinanthus
- Tedingea
- Traubia
- Ungernia
- Urceolina
- Vagaria
- Vallota
- Worsleya
- Zephyra
- Zephyranthes

Taccaceae

Classification (Bentham and Hooker)

- Phanerogams
- Monocotyledons
- Epigynae
- Taccaceae

General characters

→ Perennial herbs
→ Basal aggregation of leaves
→ Leaves - alternate, spiral, petiolate, more or less sheathing; Leaf sheaths not tubular, with free margins. Leaves usually simple, rarely compound. Lamina dissected or entire. If entire - lanceolate or ovate
→ Perianth - 6 tepals, 2 whorled, isomerous; green, purple or brown
→ Inflorescence - umbels or cymose
→ Flowers - individually bracteate (bracts long, filiform), regular, trimerous, cyclic; Pentacyclic; Perigone tube present (campanulate)
→ Androecium - 6; adnate to the perianth tube, free of one another; 2 whorled; Stamens - 6, diplostemonous; petaloid. Anthers adnate;

dehiscing via longitudinal slits, introrse,
→ Gynoecium - Tricarpellary, Partly petaloid, Carpels isomerous with perianth; Gynoecium Syncarpous; eu-syncarpous; inferior. Ovary 1 locular. Styles 1, apical; Stigmas-1; 3 lobed. Placentation Parietal; Many ovules per cavity.
→ Fruit- dehiscent or indehiscent, a capsule or a berry
→ Seeds with oly endosperm.

Genera included
→ Tacca.

Dioscoreaceae

Classification (Bentham and Hooker)

- Phanerogams
- Monocotyledons
- Epigynae
- Dioscoreaceae

General characters

→ Shrubs, herbs or lianas
→ Usually climbing rarely self supporting.
→ Mostly stem twiners or scrambling.
→ Leaves usually alternate rarely opposite, usually spiral; petiolate, sheathing to non-sheathing; usually simple, rarely compound (palmate), Lamina usually entire. Stipulate or exstipulate
→ Plants - dioecious (usually) rarely monoecious or hermaphrodite. Floral nectaries present. Nectar secretion from the gynoecium.
→ Inflorescence - Panicles, racemes or spikes
→ Flowers - bracteate, bracteolate, small, regular, trimerous, Perigone tube usually present.
→ Perianth - 6 tepals, jointed, 2 whorled; isomerous,

sepaloid or petaloid; either similar or dissimilar
- Androecium — usually 6 rarely 3, adnate to the perianth, free of one another or coherent (monadelphous). Stamens 6 or 3, isomerous with the perianth or diplostemonous. Anthers dorsifixed or adnate; dehiscing via longitudinal slits; extrorse or introrse, appendaged or unappendaged.
- Gynoecium — Tricarpellary, Syncarpous, inferior, ovary 3 locular. Styles 1 or 3, free or partially jointed; apical; stylar canal present; stigmas dorsal to the carpels. Placentation axile. Ovules 2 per locule or rarely 3-50 per locule.
- Fruit — Fleshy or non-fleshy; dehiscent or indehiscent; capsule or berry or samara.
- Seeds — oily endosperm

Genera included
- Borderea
- Dioscorea
- Epipetrum
- Rajania
- Stenomeris
- Tamus.

Series: Coronarieae

Families:

Roxburghiaceae

Liliaceae

Ponteridaceae

Philydraceae

Xyrideae

Mayacaceae

Commeliniaceae

Rapateaceae

Roxburghiaceae

→ **Classification** (Bentham and Hooker)

    Phanerogams
    Monocotyledons
    Coronarieae
    Roxburghiaceae

General Characters

→ Herbs, lianas or shrubs
→ Leaves - alternate or opposite or whorled; when alternate, distichous; herbaceous or leathery; Petiolate; non-sheathing. Lamina - entire; Margins - entire
→ Plants usually hermaphrodite or monoecious.
→ Flowers solitary or aggregated in inflorescence - cymes or racemes
→ Flowers - bracteate or ebracteate; regular, 2 merous or tetramerous; cyclic - pentacyclic or tricyclic Perigone tube may or may not be present.
→ Perianth - 4 tepals, free or joined; 2 whorled; Isomerous (2+2); sepaloid or petaloid, similar in 2 whorls
→ Androecium - 4, free or basally connate; often

Families of Monocotyledons

monadelphous; 1 or 2 whorled; Stamens - 4; diplostemonous; Anthers basifixed; dehisce via longitudinal slits; introrse.

→ Gynoecium - Bicarpellary; carpels reduced in number relative to the perianth. Syncarpous, Superior to partly inferior. Ovary unilocular; Stylate or non stylate. Basal placentation. 3-50 ovules per locule.

→ Fruit - non-fleshy, dehiscent or indehiscent; a capsule.

→ Seeds - oily endosperm

Genera included

→ Croomia

→ Stemona

→ Stichoneuron.

Liliaceae

Classification (Bentham and Hooker)

Phanerogams
Monocotyledons
Coronarieeae
Liliaceae

General Characters

→ Annual, biennial and perennial herbs and shrubs. Majority of the plants are xerophytic.
→ Roots are adventitious and fibrous.
→ Stem - underground, bulbs or rhizomes. The inflorescence may develop on a special aerial stem known as 'scape'.
→ Leaf - Simple, radical or cauline. They may be alternate or whorled. The leaves of Aloe are fleshy and succulent. The plants of Asparagus bear scale-like leaves. However, they possess phylloclades for the purpose of carbon assimilation.
→ Inflorescence - usually racemose type, but in Rohdea, the flowers are found to be arranged spirally on a spadix. Gloriosa - solitary and terminal. Gonioschypha - spikes

Families of Monocotyledons

→ **Flowers** – Pedicellate, bracteate rarely ebracteate, hermaphrodite, complete, regular, hypogynous and usually trimerous. rarely tetra or pentamerous.

→ **Perianth** – 6 perianth leaves arranged in 2 whorls each whorl consisting of 3 segments. Segments may be free or united, valvate or imbricate, petaloid or sepaloid and regular.

→ **Androecium** consists of 6 stamens arranged in 2 whorls; the stamens opposite the perianth leaves; stamens free or adnate to the perianth leaves. The filaments may be free or connate. Anthers- dithecous, introrse and basifixed.

→ **Gynoecium** – tricarpellary, syncarpous. Ovary – superior and trilocular. Axile placentation; Style single; Stigma trilobed; Numerous ovules arranged in 2 rows in each loculus.

→ **Fruit** – fleshy, berry or capsule.

→ **Seeds** – Endospermic with a straight embryo.

→ **Pollination** – entomophilous.

Genera included under Liliaceae

→ Cardiocrinum
→ Erythronium
→ Fritillaria
→ Gagea
→ Lilium
→ Lloydia
→ Medeola
→ Nomocharis
→ Notholirion
→ Tulipa

Ponteridaceae

Classification (Bentham and Hooker)

Phanerogams
Monocotyledons
Coronarieae
Ponteridaceae

General characters

→ Small family of aquatic plants floating or rooted under water.

→ Have Sympodial or rarely monopodial branches.

→ Leaves - Spiral, alternate, venation parallel, petiole occasionally inflated

→ Inflorescence - Racemose protected at the base by a spathula leaf sheath or may be reduced to a single flower.

→ Flowers - blue, purple or white, showy, usually hermaphrodite, zygomorphic

→ Perianth - 6 tepals.

→ Androecium - 6 stamens, 3 fertile and 3 sterile staminodes, epipetalous, dorsal petal usually with distinct colour compared to others.

Families of Monocotyledons

→ Nectars are present (septate)
→ Gynoecium – Syncarpellary, Superimposed ovary, unilocular or trilocular; Axile or Parietal Placentation.
→ Fruit – dry capsule
→ Seeds with abundant endosperm and can remain viable for 7 years.

Genera included under Pontederiaceae

→ Eichhornia
→ Eurystemon
→ Heteranthera
→ Hydrothrix
→ Monochoria
→ Pontederia
→ Reussia
→ Scholleropsis
→ Zosterella.

Phylidraceae

Classification (Bentham and Hooker)

    Phanerogams
    Monocotyledons
    Coronarieae
    Phyledraceae

General characters

→ Perennial herbs
→ Basal aggregation of leaves; rhizomatous or cormous.
→ Leaves - alternate, spiral and distichous (the lower distichous and upper spiral); flat, sessile, sheathing; borne edgewise to the stem; simple. Lamina entire; linear or lanceolate; parallel venation; exstipulate. Margin entire.
→ Flowers aggregated in inflorescence - spikes.
→ Flowers - bracteate (bracts rather large, flowers sometimes partially fused with them), very irregular, zygomorphic, Flowers - cyclic, tetracyclic
→ Perianth - 6 tepals, joined, 2 whorls; the two lateral lateral members of the inner whorl fused with the upper of the outer whorl to form a large,

broad upper lip, the Median member of the inner whorl forming a large lower lip, and the laterals of the outer whorl small and sometimes fused with the solitary stamen.

→ Petaloid, isomerous, white or yellow.
→ Androecium -1, free of the perianth or adnate. Filaments appendiculate or not. Anthers versatile; dehiscing via longitudinal slits; extrorse to introrse
→ Gynoecium - Tricarpellary, Syncarpous, Superior, unilocular; or 3 locular; Styles -1, attenuate from the ovary or form a depression at the top of the ovary; Stigma -1 either 1 or 3 lobed; Placentation usually axile rarely parietal. Many ovules (15-100) per locule.
→ Fruit - not fleshy; dehiscent, a capsule or a berry.
→ Seed - Endospermic with straight embryo.

Genera included under Philydraceae

→ Helmholtzia
→ Philydrum
→ Orthothylax
→ Philydrella

Xyridaceae

Classification (Bentham and Hooker)

Phanerogams
Monocotyledons
Coronarieae
Xyridaceae

General characters

→ Perennial herbs, with a basal aggregation of leaves, rhizomatous
→ Leaves - alternate, spiral or distichous; herbaceous or leathery; sessile; sheathing. Sheaths with free margins. Entire. Parallel venation
→ Plants hermaphrodite; Nectaries absent
→ Inflorescence - spikes and in heads. Ultimate inflorescence - racemose
→ Flowers - trimerous, cyclic - Tetracyclic or Pentacyclic.
→ Perianth with distinct calyx and corolla. Calyx 2·3, 1 whorl, polysepalous or partially gamosepalous. Corolla 3; 1 whorl, poly or gamopetalous, imbricate; unequal; yellow, white or blue
→ Petals when free clawed or sessile.
→ Androecium - 3 or 6, 1 or 2 whorl; Staminodes

When present 3 in outer whorl, stamens 3, isomerous with perianth; oppositiperianth. Anthers dehisce via longitudinal slits. latrorse.

→ Gynoecium - tricarpellary, syncarpous, superior, ovary 1 locule or 3 locule. Styles-1 apical, three branched; stigmas 1 or 3. placentation when unilocular parietal or basal, when trilocular axile.
Many ovules per locule;
→ Fruit - non-fleshy, dehiscent capsule. (loculicidal or circumscissile)
→ Seed - Endosperm may or may not be oily.

Genera Included
- → Abolboda
- → Achlyphila
- → Aratitryopea
- → Orectanthe
- → Xyris.

Mayacaceae

Classification (Benthany and Hooker)

- Phanerogams
- Monocotyledons
- Coronarieae
- Mayacaceae

General characters

→ Creeping herbs
→ Leaves - Submerged to emergent
→ Leaves - Small, alternate, spiral, imbricate, sessile, non-sheathing, simple. Lamina entire or dissected, linear or lanceolate
→ Flowers solitary, terminal, bracteate, small, regular, trimerous, tetracyclic, hermaphrodite, floral nectaries absent.
→ Perianth with distinct calyx and corolla. Calyx 3 in 1 whorl, polysepalous, regular, imbricate or valvate. Corolla 1 in 1 whorl, polypetalous, imbricate, regular, pale pink to purple or white. Petals shortly clawed; entire.
→ Androecium - 3, 1 whorl; isomerous with perianth, oppositipetalous; filaments slender, glabrous.

Anthers basifixed, non-versatile, dehiscing via pores to dehiscing via short slits; bilocular to 4 locular;

→ Gynoecium - tricarpellary, syncarpous, superior ovary; ovary 1 locule; sessile; Style-1, attenuate from the ovary, apical; Stylar canal present; Stigmas 1 or 3; when single, 1 or 3 lobed (capitate to trifid). Parietal placentation. 25-100 ovules per locule.

→ Fruit - non-fleshy, dehiscent, capsule.

→ Seeds - endospermic (not oily)

Genera

→ Mayaca

Commelinaceae

Classification (Bentham and Hooker)

Phanerogams
Monocotyledons
Coronarieae
Commelinaceae.

General characters

→ Succulent perennial or annual herbs having nodose stems.

→ Roots fibrous or sometimes much thickened and tuber like. Climbing plants are rare (Streptolirion). Only epiphyte is Cochliostema.

→ Leaves – alternate, flat or trough like, entire with parallel venation, sheathing by a basal membranous and often closed sheath.

→ Flowers – usually actinomorphic, rarely zygomorphic, hermaphrodite, trimerous, hypogynous. In *commelina benghalensis* small cleistogamic flowers are found on subterranean branches of rhizome.

→ Perianth consists of 2 series. Outer series consists of 3 free, green sepal like tepals, imbricate (rarely gamotepalous and rarely the tepals petaloid); the inner series consists of 3 petaloid (coloured) tepals, free, equal or unequal, rarely united into a tubular structure, the third tepal sometimes much reduced.

→ Androecium consists of 6 stamens, due to abortion the fertile stamen is reduced to 3.; dithecous; filaments distinct and rarely connate, often bearded with coloured moniliform hairs

→ Gynoecium consists of 3 carpels, syncarpous, ovary superior, trilocular, the ovules 1- few per locule and orthotropous, placentation axile, style 1, stigma 1 and capitate or trifid.

→ Fruit - loculicidal capsule, sometimes enclosed by fleshy outer perianth segments.

→ Seeds- endospermic with copious and mealy endosperm

→ Pollination - entomophilous.

Genera included under commelinaceae

- Aetheolirion
- Amischotolpe
- Aneilema
- Anthericopsis
- Belosynapsis
- Buforrestia
- Callisia
- Cochliostema
- Coleotrype
- Commelina
- Cyanotis
- Dichorisandra
- Dictyospermum
- Elasis
- Floscopa
- Geogenanthus
- Gibasis
- Gibasoides
- Matudanthus
- Murdannia
- Palisota
- Pollia
- Polyspatha
- Porandra
- Pseudoparis
- Rhoeo
- Rhopalephora
- Sauvallea
- Siderasis
- Spatholirion
- Stanfieldiella
- Streptolirion
- Thyrsanthemum
- Tinantia
- Tradescantia
- Tricarpelema
- Tripogandra
- Weldenia

Rapateaceae

Classification (Bentham and Hooker)

- Phanerogams
- Monocotyledons
- Coronarieae
- Rapateaceae

General characters

→ Perianth in 2 series
→ 6 anthers
→ Numerous capitate flowers subtended by two foliaceous bracts

Genera

→ Amphiphyllum
→ Cephalostemon
→ Duckea
→ Epidryos
→ Guacamaya
→ Kunhardtia
→ Marahuacaea
→ Maschalocephalus
→ Monotrema
→ Phelpsiella

→ Potarophytum
→ Rapatea
→ Saxo-fridericia
→ Schoenocephalium
→ Spathanthus
→ Stegolepis
→ Windsorina

Series: Calycineae

Families:

Flagellariaceae

Juncaceae

Palmae

Flagellariaceae

Classification (Bentham and Hooker)

- Phanerogams
- Monocotyledons
- Calycineae
- Flagellariaceae

General Characters

→ Lianas; with dichotomous branching from sympodial rhizomes; Perennial.

→ High climbing (stem - cane like); tendril climbers

→ Leaves - alternate; distichous; petiolate; sheathing. Leaf sheath with joined margins. Leaves - simple; lamina - entire, lanceolate; Parallel venation.

→ Plants hermaphrodite; lack nectaries.

→ Flowers aggregated in inflorescence - Panicles, terminal; bracteate, trimerous, Pentacyclic.

→ Perianth represented by tepals - 6 free; 2 whorled of 3 each; isomerous; petaloid; white

→ Androecium – 6 stamens – Members free from perianth, free from one another; Present in 2 whorl. Anthers – sagittate basifixed;

→ Gynoecium – tricarpellary, syncarpous, superior Carpels isomerous with Perianth; Ovary 2 locular Styles 3 – Mostly free rarely connate basally; Stigmas – 3; Axile placentation; Ovule 1 per locule.

→ Fruit – fleshy, indehiscent, drupe
→ Seeds with endosperm (not oily)

Genera
→ Flagellaria.

Juncaceae

Classification (Bentham and Hooker)
- Phanerogams
- Monocotyledons
- Calycineae
- Juncaceae

General characters
→ Commonly known as "rush family" as they are Rush like perennial or annual herbs rarely under shrubs.

→ Adventitious hairy fibrous roots

→ Stem – erect or horizontal rhizome creeping sympodial rhizome, short.

→ Leaves – basal, tufted, linear or filiform, grasslike and flat or terete; sheathing basally; sheath may either be opened or closed

→ Inflorescence – solitary or in panicle, corymb or head like cyme

→ Flowers – hermaphrodite, rarely unisexual;

if unisexual plants are dioecious, hypogynous actinomorphic, trimerous.

→ Perianth - tepals 6 rarely 3; arranged in 2 whorls; glumaceous, scarious or coriaceous; rarely petaloid green, white, yellow or black, free, imbricate aestivation; persistent

→ Androecium - stamens 6; 2 whorls; anthers bithecous; filament triangular very slender; short or long, basifixed, introrse

→ Gynoecium - Tricarpellary, syncarpous, superior, trilocular with axile placentation; rarely unilocular with parietal placentation; style long or absent ending in 3 palmose stigmas; ovules many, anatropous

→ Fruit - loculicidal capsule

→ Seed - small, endospermic with membranous testa

Genera
→ Andesia
→ Distichia
→ Juncus
→ Luzula
→ Marsippospermum
→ Oxychloe
→ Prionium
→ Rostkovia

Families of Monocotyledons

Palmae (Palmaceae / Arecaceae)

Classification (Bentham and Hooker)
Phanerogams
Monocotyledons
Calycinae
Palmae.

General characters

→ Adult palms have a tall, woody, unbranched stem bearing a crown of leaves and have its circumference ring marked with the bases of leaves which have perished. Some are climbers and possess slender stems known as Rattans or cane-palms.

→ Roots adventitious.

→ Stem - reaches a height of 60 ft or more than that and remains covered with remnants of persistent leaf bases.

→ Phytelephus possesses a very short and thick stem and the leaves.

→ Some of the palms may have a height of more than 150 ft

→ Branching is rare. However it is found in Hyphaene thebaica

- Leaves – few in number and often very large 2 types – Palmately divided leaves and Pinnately divided leaves. commonly called as fan palm and feather palm types.
- In fan palms (Borassus) – leaf blade is entire and folded in bud and as the leaf expands the folds become incised from the margin inwards. The same is found in feather palms.
- Leaf has a large, stout petiole with a broad sheathing base.
- Iriartea ?, thorns occur on stem, leaves and even roots. Stem thorns may also be present within the leaf sheath.
- Inflorescence – huge.

  Some palms are monocarpic. The growth of the plant stops at a particular time and develops a huge inflorescence. This production exhausts the food reserves of the plant to such extent that it dies after fruiting (Corypha)

  Usually the palms are polycarpic (i.e) bear axillary inflorescence which wither away after fruiting (Borassus)
- Inflorescence – simple (Borassus), compound spike (Cocos) and profusely branched panicle (Daemonorops)

Families of Monocotyledons

→ young inflorescence remain enclosed in a spathe (spadix)

→ Flowers – unisexual, actinomorphic, incomplete, hypogynous. Usually sessile and rarely bears short pedicels. Very small in size and present in very large numbers in inflorescence.

→ Usually monoecious (cocos); rarely dioecious (Phoenix). In monoecious, female flowers occur at the base of the branches whereas the male ones are crowded in the upper part.

→ Perianth – 6 perianth leaves (tepals) arranged in 2 whorls of 3 each. Polyphyllous and persistent. In *Areca catechu*, the perianth leaves of the outer whorl are much smaller than the perianth leaves of the inner whorl. They are green to yellow or white in colour.

→ Androecium – 6 stamens arranged in 2 whorls of 3 each. Rarely reduced to 1 or increased to infinity. They are polyandrous. Anthers – bilobed. Filaments – short and distinct.

→ Gynoecium – tricarpellary, apocarpous or syncarpous. Ovary – superior and trilocular, rarely unilocular. Each carpel has a single anatropous ovule. Style short

→ Fruit – berry with a fleshy or fibrous covering. (*Phoenix*). In *cocos*, the fruit is a fibrous drupe.
→ Seeds vary in shape and size. Endospermic.

Families of Monocotyledons

Genera included under Palmaceae

- Archontophoenix
- Areca
- Astrocaryum
- Attalea
- Bactris
- Beccariophoenix
- Bismarckia
- Borassus
- Butia
- Calamus
- Ceroxylon
- Cocos
- Coccothrinax
- Copernicia
- Corypha
- Elaeis
- Euterpe
- Hyphaene
- Jubaea
- Latania
- Licuala
- Livistona
- Mauritia
- Metroxylon
- Nypa
- Parajubaea
- Phoenix
- Pritchardia
- Raphia
- Rhapidophyllum
- Rhapis
- Roystonea
- Sabal
- Salacca
- Syagrus
- Thrinax
- Trachycarpus
- Trithrinax
- Veitchia
- Washingtonia

Series: Nudiflorae

Families:

Pandaneae

Cyclanthaceae

Typhaceae

Aroideae

Lemnaceae

Pandaneae

Classification (Bentham and Hooker)

Phanerogams
Monocotyledons
Nudiflorae
Pandaneae (Pandanaceae)

General Characters

→ Plants palm like; branched shrubs or trees. Supported by silt roots
→ Stout aerial prop roots; rarely root climbers
→ Stem — aerial, erect or bent, branched, stout
→ Leaf — long, linear, sessile, serrate spinous margin, often forming crown at the apex of stem, tough, fibrous and leathery.
→ Inflorescence — Paniculate or spikes covered by spathaceous or foliaceous bracts.
→ Flower — unisexual, lack perianth, plants dioecious.
→ Androecium — numerous stamens, densely packed or separated or in fasciculate clusters;

scattered over a spadix like axis; filament-distinct or connate; anthers bithecous, basifixed
→ Gynoecium absent in male flowers. In female flowers – staminodes either small or absent. Pistil numerous either isolated or coherent into bundles; ovary superior, unilocular, ovules 1-many, basal or parietal placentation, ovules anatropous; style short or absent; stigma 1.
→ Fruit – syncarpous (multiple of berries or drupes)
→ Seed – small with copious oily endosperm
→ Pollination – entomophilous or anemophilous

Genera
  → Freycinetia
  → Pandanus
  → Sararanga

Cyclanthaceae

Classification (Bentham and Hooker)

    Phanerogams
        Monocotyledons
            Nudiflorae
                Cyclanthaceae

General Characters

→ Shrubs or lianas or herbs; perennial; with a basal aggregation of leaves; self supporting rarely epiphytic or climbing.

→ Leaves - alternate, spiral rarely distichous, petiolate, simple or compound (palmate or bifoliate); lamina - entire or dissected (usually); one veined or palmately veined.

→ Plants monoecious; flowers - unisexual. Female flowers with staminodes and male flowers lack gynoecium.

→ Inflorescence - terminal or axillary; pedunculate spadices; spatheate. Each spadix subtended by 2-8 spathes; flowers - small.

→ Perianth — tepals 4 or vestigial or absent If 4, free or joined; 1 whorled, sepaloid, fleshy or non fleshy; Persistent; accrescent

→ Androecium 10-150; free of Perianth; free from one another; triplostemonous to polystemonous; often with basally bulbous filaments; Anthers basifixed.

→ Gynoecium - 4 carpelled; carpels isomerous with Perianth; Gynoecium - Syncarpous, Partly inferior or completely inferior. Ovary 1 locular; Styles 1-4; free to partially joined or absent in some cases; Stigmas - 4; laterally compressed or flat, broad and fleshy

→ Fruit - fleshy, indehiscent; a berry.

→ Seeds copiously endospermic

Genera
- → Asplundia
- → Carludovica
- → Chorigyne
- → Cyclanthus

→ Dicranopygium
→ Evodianthus
→ Ludovia
→ Schultesiophytum
→ Sphaeradenia

→ Stelestylis
→ Thoracocarpus

→ Dianthoveus

Typhaceae

Classification (Bentham and Hooker)

Phanerogams
Monocotyledons
Nudiflorae
Typhaceae

General characters

→ Water or Marsh herbs, with a perennial creeping rhizome.

→ Rhizome bears erect simple shoots and distichous, long, narrow, entire sheathing leaves. Rhizome also bears two lateral rows of scales. Branches may spring out from the axils of the scales.

→ Branches become thickened at their tips to form a knee like band.

→ Inflorescence – unisexual flowers on dense spike or cylindrical spadix, the male flowers in the upper portion and female flowers in the lower portion. Male flowers arise directly from the flattened axis of the spike. Female flowers are usually borne on crowded, short cylindrical outgrowth.

→ Flowers are subtended by one caducous bractlike spathe.

→ Male flowers → 1-5 stamens; monadelphous; filaments being connate and bear long silky hairs. Only 1 stamen is present in some cases (Typha minima). The hairs or scales developing at the base of the flowers on the axis of the spike surround the flowers.

→ Female flowers :— generally remain subtended by small, spathulate scale like bracts. Bracts are absent in some (T. latifolia). The normal flower possesses a long gynophore on which many irregularly arranged simple hairs have been found. Style is slender which terminates in a unilateral elongated stigma. Ovary - superior, unilocular, with a single anatropous pendulous ovule; Placentation - parietal.

→ Fruits — achene, covered with a membranous or leathery pericarp

→ Seeds — small, endospermic with a straight embryo.

→ Pollination — anemophilous

Genera included under Typhaceae

Typha

Families of Monocotyledons

Araceae

Classification (Bentham and Hooker)

- Phanerogams
- Monocotyledons
- Nudiflorae
- Aroideae

General characters

→ Plants may be small or large herbs, sometimes shrubs with sympodial branches climbing by aerial roots (pothos). Some are epiphytic and climbing, some are hydrophytes and some prefer marshy places.

→ Most plants contain watery or milky juice (latex). Sometimes the sap is acrid and with pungent odour. Resin canals and calcium oxalate crystals are also present within the tissues.

→ Root - adventitious, sometimes epiphytic plants produce aerial roots. Usually the velamen layer is found on the aerial roots to absorb the moisture from atmosphere. In epiphytic climbing plants, the roots serve both the purpose of absorption as well as of clasping. The clasping roots are negatively phototrophic and for the plant to its support.

→ Stem - Erect or climbing; underground stems are usually developed in the form of root-stocks, tubers or corms, when often have a very pungent taste

→ Leaves - large, simple or compound, usually with parallel venation, petiolate, petiole with a sheathing base. Cordate, sagittate or hastate, lamina entire, lobed, pinnate or perforate; Arrangement may be alternate, distichous or spiral. In herbaceous species - Leaves are few, clustered or solitary, radical.

→ Inflorescence - small, sessile, unisexual rarely bisexual flowers arranged on a spadix which is more or less completely enclosed in a green or coloured leaf like bract (spathe). When flowers are unisexual, they are usually monoecious with the male flowers towards the apex of the spadix and females below. Racemose, spadix with spathe. The entire inflorescence may be covered with flower or this ends in a sterile conical portion.

Families of Monocotyledons

→ Flowers – sessile, small, actinomorphic, regular, bi to trimerous or dimerous, unisexual, rarely hermaphrodite, monoecious or dioecious. In monoecious condition, the male flowers are always towards the apex of the spadix and the female flowers below. A neutral flower zone separates the male and female flowers.

→ Perianth – absent or 4-6 scaly structures present, free or connate segments, rarely cupular.

→ Androecium – 2,4,6 or 1, usually 6, hypogynous, distinct or confluent; anthers two-celled occasionally four-celled, free or connate by means of thickened connective, anthers usually dehisce by terminal pores. Staminodes represent the stamens in female flowers. The stamens may be arranged in 1 or 2 whorls. When arranged in one whorl the filaments are fused at the base forming a synandrium.

→ Gynoecium – one, two to many carpels; ovary superior rarely inferior, one to many loculed; the ovules one to many in each loculus. Placentation may be basal, axile or apical; style short; stigma disc like or lobed.

→ Fruit – berry, sometimes many small, free or connate berries or drupes adnate to spadix.
→ Seeds – Endospermic in each berry or drupe, one to many seeds, rarely non-endospermic.
→ Pollination – entomophilous.

Genera included under Araceae

- → Aglaodorum
- → Aglaonema
- → Alloschemone
- → Alocasia
- → Ambrosina
- → Amorphophallus
- → Amydrium
- → Anadendrum
- → Anaphyllopsis
- → Anaphyllum
- → Anchomanes
- → Anthurium
- → Anubias
- → Aridarum
- → Ariopsis
- → Arisaema
- → Arisarum
- → Arophyton
- → Arum
- → Asterostigma
- → Biarum
- → Bognera
- → Bucephalandra
- → Caladium
- → Calla
- → Callopsis
- → Carlephyton
- → Cercestis
- → Chlorospatha
- → Colletogyne
- → Colocasia
- → Cryptocoryne
- → Culcasia
- → Cyrtosperma
- → Dieffenbachia
- → Dracontioides
- → Dracontium
- → Dracunculus
- → Eminium
- → Epipremnum
- → Filarum
- → Furtodoa
- → Gearum
- → Gonatanthus
- → Hapaline
- → Helicodiceros
- → Heteroaridarum
- → Heteropsis
- → Holochlamys
- → Homalomena
- → Hottarum
- → Jasarum
- → Lagenandra
- → Lasia
- → Lasimorpha
- → Lysichiton
- → Mangonia
- → Monstera
- → Montrichardia

Families of Monocotyledons

Araceae contd...

- Nephthytis
- Orontium
- Pedicellarum
- Peltandra
- Philodendron
- Phymatarum
- Pinellia
- Peptospatha
- Pistia
- Podolasia
- Pothoidium
- Pothos
- Protarum
- Pseudodracontium
- Pseudohydrosme
- Pycnospatha
- Remusatia
- Raphidophora
- Rhodospatha
- Sauromatum
- Scaphispatha
- Schismatoglottis
- Scindapsus
- Spathantheum
- Spathicarpa
- Spathiphyllum
- Stenospermation
- Steudnera
- Stylochaeton
- Symplocarpus
- Synandrospadix
- Syngonium
- Taccarum
- Theriophonum
- Typhonium
- Typhonodorum
- Ulearum
- Urospatha
- Urospathella
- Xanthosoma
- Zamioculcas
- Zantedeschia
- Zomicarpa
- Zomicarpella

Lemnaceae

Classification (Bentham and Hooker)

    Phanerogams
    Monocotyledons
    Nudiflorae
    Lemnaceae

General Characters

→ Smallest and least differentiated angiosperms of the world. They are aquatic in nature found floating in various fresh waters.

→ Plant body consists of green dorsiventral, scale-like shoots. Plants lack green leaves and the flat green shoot performs the functions of leaf. From the ventral surface of the flattened stem one or several adventitious roots come out. The apex of each root is converted by a few layered sheath (Root cap) which is visible to the unaided eyes. Roots are absent in _Wolffia_

→ Inflorescence — quite simple, in _Lemna_ and _Spirodela_ it arises in a pocket

→ Flowers — unisexual, Plants — Monoecious (i.e) both male and female flowers develop in the same inflorescence. Flowers are naked (ie) without perianth. Male flower consists of single stamen. Filament is stout bearing at its apex a pair of dithecous anther halves. Pollen grains are spherical and covered with small warty outgrowths. Female flower consists of single carpel. Pistil is flask shaped with a short funnel shaped stigma. Ovary is unilocular with one to 6 basal, erect ovules.

→ Seed possess a thick fleshy outer and a thin inner coat. Embryo consists of large cotyledon surrounded by scanty endosperm.

→ Pollination by wind, water or animals.

Genera included under Lemnaceae

→ Lemna
→ Spirodela
→ Pseudowolffia
→ Wolffia
→ Wolffiella
→ Wolffiopsis.

SERIES: APOCARPAE

FAMILIES:

Triurideae

Alismaceae

Najadaceae

Triuridaeae (Triuridaceae)

Classification (Bentham and Hooker)

- Phanerogams
- Monocotyledons
- Apocarpae
- Triuridaeae

General Characters

→ Achlorophyllous; Pallid or purplish herbs.
→ Leaves much reduced. Parasitic on the roots of the host. Leaves — Minute or small, alternate, Membranous, sessile, non-sheathing, simple. Lamina entire
→ Plants monoecious or dioecious rarely hermaphrodite
→ Flowers in inflorescence - racemes, cymes.
→ Flowers- bracteate, small, regular, tetracyclic floral receptacle developing an androphore or a gynophore. Perigone tube present and is short, lobes often reflexed
→ Perianth - 3-10 tepals, joined, 1 whorled; Petaloid; without spots; white or red to Purple or hyaline
→ Androecium 2-6 stamens; free of one another, free of Perianth, all equal; 1 whorled;

Stamens 2-6, alterniperianthial or oppositiperianthii filantherous or with sessile anthers; Anthers - non versatile, extrorse.

→ Gynoecium 6-50 carpelled; carpels isomerous with the perianth, apocarpous; euapocarpous, superior; carpel stylate; 1 ovuled, Placentation basal, stigma present.

→ Fruit - non fleshy, aggregate.

→ Seeds - Endospermic.

Genera
- Andruris
- Hexuris
- Hyalisma
- Sciaphila
- Seychellaris
- Soridium
- Triuris

Alismaceae (Alismataceae)

Classification (Bentham and Hooker)

- Phanerogams
- Monocotyledons
- Apocarpae
- Alismaceae

General characters

→ Mostly annual or perennial marsh or aquatic herbs.
→ Fibrous roots develop from stout rhizomes.
→ Plants may be cauline and erect or with leaves floating.
→ Leaves – basal, long petioled, sheathing basally blades vary from linear to ovate or the bases hastate or sagittate. Parallel venation
→ Inflorescence – Raceme or Panicle
→ Flower – Pedicellate, often in whorls, bracteate, bisexual or unisexual, complete or incomplete, actinomorphic, trimerous, hypogynous.
→ Perianth – 4 tepals, polytepalous (free segments), tepals arranged in 2 series, regular, imbricate, outer 3 tepals and sepal like, herbaceous

and persistent, the inner three larger and petaloid segments are deciduous.

→ Androecium – stamens 6, or more, rarely 3, free, anthers dithecous, extrorse or dehiscing by lateral slits. In *Alisma*, there is a single whorl of 6 stamens. In *Sagittaria*, the outer whorl of 6 stamens is followed by numerous stamens. In *Hydrocleis* and *Lemnocharis*, the fertile stamens remain surrounded by sterile stamens or staminodes.

→ Gynoecium consists of 6 to many free carpels, spirally arranged or in a single whorl. *Sagittaria*, the number of carpels is indefinite; receptacle is flat or convex; ovary superior, unilocular, ovules are solitary or many, basal, the style mostly needle like and persistent, the stigma simple and 1 and hardly may be distinguished from style.

→ Fruit – achene, rarely follicular, when follicle it dehisces basally.

→ Seed – non-endospermic, with a straight or curved embryo.

→ Pollination – Entomophilous.

Genera included under Alismaceae

- Albidella
- Alisma
- Alismaticarpum
- Astonia
- Baldellia
- Burnatia
- Butomopsis
- Caldesia
- Damasonium
- Echinodorus
- Helanthium
- Hydrocleys
- Limnocharis
- Limnophyton
- Luronium
- Ranalisma
- Sagisma
- Sagittaria
- Wiesneria.

Najadaceae

Classification (Bentham and Hooker)

    Phanerogams
    Monocotyledons
    Apocarpae
    Najadaceae

General characters

→ Annual or perennial herbs
→ Hydrophytic; non marine; rooted. Leaves submerged.
→ Leaves – opposite to whorled or alternate; sessile; sheaths present, with free margin; leaves – simple, lamina – entire, linear. Lamina margins serrate to dentate or entire.
→ Unisexual flowers present. Plants monoecious or dioecious
→ Perianth of tepals or vestigial or absent; free or joined; sepaloid.
→ Androecium – 1, stamen with sessile anthers;
→ Gynoecium – 1 carpel; superior. Carpel 1 ovuled, basal placentation, anatropous.
→ Fruit – non-fleshy, indehiscent, achene

→ seeds with starch and non-endospermic

Genera

→ Najas

Series: Glumaceae

Families:

Eriocauleae

Centrolepideae

Restiaceae

Cyperaceae

Gramineae

Families of Monocotyledons

Classification — Eriocauleae (Bentham and Hooker)

- Phanerogams
- Monocotyledons
- Glumaceae
- Eriocauleae

General characters

→ Perennial herbs with basal aggregation of leaves.
→ Leaves - alternate, usually spiral, rarely distichous;
→ Unisexual, Plants usually monoecious or rarely dioecious. Female flowers lack staminodes. Gynoecium of male flowers are pistillodeal to vestigial
→ Inflorescence - heads, racemose
→ Flowers - 2 or 3 merous, cyclic. Floral receptacle developing on androphore or may lack both androphore and gynophore. Perigone tube present
→ Perianth with distinct calyx and corolla or sepaline. 4-6 joined, 2 whorled or rarely 1 whorled

Calyx 2 or 3 in 1 whorl; polysepalous or gamosepalous; bilabiate or regular. Corolla 2 or 3 in 1 whorl; poly or gamopetalous; unequal or regular.

→ Androecium 2-6; di or trimerous; adnate to corolla; free of one another; 1 or 2 whorled; Anthers basifixed, introrse; isomerous with Perianth or diplostemonous (Stamens)

→ Gynoecium - 2 or 3 carpelled. Syncarpous; Synstylovarious, superior; ovary 2 or 3 locular; Style 1-3; free to partially joined, apical; Stigmas 2 or 3; Placentation apical, orthotropous ovule

→ Fruit - not fleshy, dehiscent, loculicidal capsule
→ Seeds - Endospermic.

Genera
→ Actinocephalus
→ Blastocaulon
→ Eriocaulon
→ Lachnocaulon
→ Leiothrix
→ Mesanthemum
→ Paepalanthus
→ Philodice
→ Rhodonanthus
→ Syngonanthus
→ Tonina.

Centrolepideae

Classification (Bentham and Hooker)

Phanerogams
Monocotyledons
Glumaceae
Centrolepideae

General characters

- Tufted plants
- Narrow leaves from base
- Flowers in highly condensed inflorescences enclosed between a pair of bracts that often have leaf like points
- Anemophilous (Pollination)

Genera

- Aphelia
- Centrolepis
- Gaimardia

Restiaceae (Restionaceae)

Classification (Bentham and Hooker)

Phanerogams
Monocotyledons
Glumaceae
Restiaceae

General characters
- Annual or Perennial rush like flowering plant
- tufted or rhizomatous herbaceous plants
- Bamboo like in overall appearance
- Stems and leaves have been reduced to sheaths
- Inflorescence - spikelets
- Flowers - extremely small, dioecious (ie) male and female flowers on separate plants

Genera included under Restiaceae
- Alexgeorgea
- Anarthria
- Anthochortus
- Apodasmia
- Askidiosperma
- Baloskion
- Calorophus
- Cannomois
- Ceratocaryum
- Chaetanthus
- Chordifex
- Coleocarya
- Dapsilanthus
- Dielsia

- → Elegia
- → Empodisma
- → Eurychorda
- → Harperia
- → Hopkinsia
- → Hydrophilus
- → Hypodiscus
- → Hypolaena
- → Lepidobolus
- → Leptocarpus
- → Lepyrodia
- → Loxocarya
- → Lyginia
- → Mastersiella
- → Meeboldina
- → Megalotheca
- → Nevillea
- → Onychosepalum
- → Platycaulos
- → Restio
- → Rhodocoma
- → Sporadanthus
- → Staberoha
- → Thamnochortus
- → Willdenowia
- → Winifredia.

Cyperaceae

Classification (Bentham and Hooker)

    Phanerogams
    Monocotyledons
    Glumaceae
    Cyperaceae

General characters

→ Sedges are usually perennial herbs and rush-like in appearance. They perennate by means of creeping sympodial rhizomes. Some are annual herbs.

→ Aerial shoots are given out from their underground rhizomes either in solitary or in clusters.

→ Stem — solid and triangular. Elongated runners are found in many sedges.

→ Leaves — 3 rows. They possess narrow blade and closed sheath.

→ Inflorescence — spike of spikelets or panicle. A single flower develops in the axil of the bract i.e. glume. In certain cases the spike like cymes are so aggregated that they look like spikes or heads.

Rarely terminal and solitary flowers are found (*oreobolus*).

→ Flower – sessile, bracteate, hermaphrodite or unisexual. Flowers borne in the axil of glume. Female flower of *carex* is borne in the axil of second glume known as utricle. The utricle forms a sac like structure around the ovary.

→ Perianth – *Eriophorum gracile*, it is represented by hairs. Some (*Scirpus lacustris*), perianth represented by bristles. Sometimes perianth represented by scales. Perianth of *oreobolus* consists of 6 tepals arranged in 2 whorls of 3 each.

→ Androecium – Rarely 6 stamens in 2 whorls of 3 each. Usually it consists of only 3 stamens of outer whorl. Sometimes the number of stamens is reduced to only 1 (*Hemicarpha*). Anthers – bilobed and basifixed. Filaments are free.

→ Gynoecium – tricarpellary or bicarpellary, syncarpous. As many feathery stigmas are found as many carpels are present. Base of style is thick and beak like. Ovary – superior, unilocular; single basal anatropous ovule. Basal placentation is believed to be derived from free central placentation.

→ Fruit – achene
→ Seed – Albuminous with small embryo.
→ Pollination – anemophilous.

Genera included under cyperaceae

- Abildgaardia
- Actinlus
- Actinoschoenus
- Afrotrilepis
- Alinula
- Androtrichum
- Anosporum
- Arthrostylis
- Ascolepis
- Ascopholis
- Baeothryon
- Baumea
- Becquerelia
- Bisboeckelera
- Blysmopsis
- Blysmus
- Bolboschoenus
- Bulbostylis
- Calyptrocarya
- Capitularina
- Carex
- Carpha
- Caustis
- Cephalocarpus
- Chorizandra
- Chrysitrix
- Cladium
- Coleochloa
- Costularia
- Courtoisina
- Crosslandia
- Cyathochaeta
- Cyathocoma
- Cymophyllus
- Cyperus
- Desmoschoenus
- Didymiandrum
- Diplacrum
- Diplasia
- Dulechium
- Egleria
- Eleocharis
- Eleogiton
- Epischoenus
- Eriophoropsis
- Eriophorum
- Erioscirpus
- Evandra
- Everardia
- Exocarya
- Exochogyne
- Ficinia
- Fimbristylis
- Fuirena
- Gahnia
- Gymnoschoenus
- Hellmuthia
- Hemicarpha
- Hymenochaeta
- Hypolytrum
- Isolepis
- Kobresia
- Kyllinga
- Kyllingiella
- Lagenocarpus
- Lepidosperma
- Lepironia

- → Lipocarpha
- → Lophoschoenus
- → Machaerina
- → Mapania
- → Mapaniopsis
- → Mariscus
- → Mesomelaena
- → Microdracoides
- → Micropapyrus
- → Monandrus
- → Morelotia
- → Neesenbeckia
- → Nemum
- → Nelmesia
- → Oreobolopsis
- → Oreobolus
- → Oxycaryum
- → Paramapania
- → Phylloscirpus
- → Pleurostachys
- → Principina
- → Pseudoschoenus
- → Ptilanthelium
- → Pycreus
- → Queenslandiella
- → Reedia
- → Remirea
- → Rhynchocladium
- → Rhynchospora
- → Rikliella
- → Schoenoplectus
- → Schoenoxiphium
- → Schoenoides
- → Schoenus
- → Scirpodendron
- → Scirpoides
- → Scirpus
- → Scleria
- → Sphaerocyperus
- → Sumatroscirpus
- → Syntrinema
- → Tetraria
- → Tetrariopsis
- → Thoracostachyum
- → Torulinium
- → Trachystylis
- → Trianoptiles
- → Trichoschoenus
- → Tricostularia
- → Trilepis
- → Tylocarya
- → Uncinia
- → Vesicarex
- → Volkiella
- → Websteria

Gramineae (Poaceae)

Classification (Bentham and Hooker)

    Phanerogams
    Monocotyledons
    Glumaceae
    Gramineae.

General characters:

→ Annual, biennial or perennial herbs or shrubs. Largest woody species are the bamboos, which may grow to more than 100 feet in height and several inches in diameter.

→ Root - Generally adventitious, fibrous, fascicled, stilt.

→ Stem - erect, prostrate or creeping. Usually hollow (fistular), rarely solid (Zea). The nodes are conspicuous and internodes are long. In many grasses the runners and suckers are also developed. Among perennial grasses, usually the rhizomes and root stocks are formed, which help in vegetative propagation. In many species tubers and corms are formed.

→ The aerial stem of this family is commonly known as 'culm' which is usually cylindrical

but in few cases flattened
- In most grasses, the main axis only develops lateral branches from the basal buds. These branches are 'tillers'.
- Leaves - foliage leaf easily divided into 2 parts, sheath and blade (lamina). Leaf sheath covers the internode partially or completely. Inner surface of the leaf sheath is glabrous (smooth) which the upper surface is grooved and either glabrous or hairy.
- At the junction of the leaf sheath and leaf blade, there occurs a thin membranous outgrowth, the ligule. The leaf blade is long, narrow, acuminate and with parallel venation.
- Usually sessile, rarely petiolate; alternate, lanceolate, entire or hairy margins, surface smooth or glabrous.
- Inflorescence - somewhat complex one to several spikelets which are combined in various manners on the main axis called rachis. Compound spikes (wheat), racemes (Festuca), panicles (Avena). Each spikelet may bear 1 to many florets attached to a central stalk called rachilla.
- Flowers - sessile, bracteate (represented by lemma

and palea), hermaphrodite or unisexual, zygomorphic, hypogynous, irregular

> Lemma is the lower and outer bract of a floret. Usually the lemma bears a long awn as an extension of the midrib at the apex or back. The floral parts borne in the axil of lemma. Palea often with two longitudinal ridges (keels or nerves), stands between the lemma and rachilla.

→ **Perianth** - sometimes absent or represented by 2 or rarely 3 minute, membranous, scale like lodicules (here the perianth leaves are known as lodicules)

→ **Androecium** - 3 stamens, rarely 6 (*Oryza*) and sometimes reduced to 2 or 1. Filaments - free, long slender and anthers - versatile. Anthers 2 celled, dehisce longitudinally.

→ **Gynoecium** - monocarpellary (though the pistil is tricarpellary but only one carpel is functional) syncarpous, unilocular ovary, superior, ovule 1, anatropous; style short, usually 2, stigma usually 2 arise from the carpellary wall, feathery. In *maize*, the

Style is long and spiky. Placentation basal
- → Fruit – usually caryopsis. In _Dendrocalamus_ it is a nut
- → Seeds – albuminous.
- → Pollination – usually anemophilous; cross pollination may also occur.

Families of Monocotyledons

Genera included under Poaceae

- Acamptoclades
- Achlaena
- Achnatherum
- Aciachne
- Acidosasa
- Acostia
- Acrachne
- Acritochaete
- Acroceras
- Actinocladum
- Aegilops
- Aegopogon
- Aeluropus
- Afrotrichloris
- Agenium
- Agnesia
- Agropyron
- Agropyropsis
- Agrostis
- Aira
- Airopsis
- Alexfloydia
- Alloeochaete
- Allolepis
- Alloteropsis
- Alopecurus
- Alvimia
- Amblyopyrum
- Ammochloa
- Ammophila
- Ampelodesmos
- Amphibromus
- Amphicarpum
- Amphipogon
- Anadelphia
- Ancistrachne
- Ancistragrostis
- Andropogon
- Andropterum
- Anemanthele
- Aniselytron
- Anisopogon
- Anomochloa
- Anthaenantiopsis
- Anthenantia
- Anthephora
- Anthochloa
- Anthoxanthum
- Antinoria
- Apera
- Aphanelytrum
- Apluda
- Apochiton
- Apoclada
- Apocopis
- Arberella
- Arctagrostis
- Arctophila
- Aristida
- Arrhenatherum
- Arthragrostis
- Arthraxon
- Arthropogon
- Arthrostylidium
- Arundinaria
- Arundinella
- Arundo
- Arundoclaytoni
- Asthenochloa
- Astrebla

- Athroostachys
- Atractantha
- Aulonemia
- Australopyrum
- Austrochloris
- Austrodanthonia
- Austrofestuca
- Austrostipa
- Avellinia
- Avena
- Axonopus
- Bambusa
- Baptorhachis
- Bealia
- Beckeropsis
- Beckmannia
- Bellardiochloa
- Bewsia
- Bhidea
- Blepharidachne
- Blepharoneuron
- Boissiera
- Boivinella
- Borinda
- Bothriochloa
- Bouteloua
- Brachiaria
- Brachyachne
- Brachychloa
- Brachyelytrum
- Brachypodium
- Briza
- Bromuniola
- Bromus
- Brylkinia
- Buchloe
- Buchlomimus
- Buergersiochloa
- Calamagrostis
- Calamovilfa
- Calderonella
- Calosteca
- Calyptochloa
- Camusiella
- Capillepedium
- Castellia
- Catabrosa
- Catabrosella
- Catalepis
- Catapodium
- Cathestechum
- Cenchrus
- Centotheca
- Centochloa
- Centropodia
- Cephalostachyum
- Chaboissaea
- Chaetium
- Chaetobromus
- Chaetopoa
- Chaetopogon
- Chaetostichum
- Chamaeraphis
- Chandrasekharani
- Chasechloa
- Chasmanthium
- Chasmopodium
- Chevalierella
- Chikusichloa
- Chimonobambusa
- Chionachne
- Chionochloa
- Chloachne
- Chloris
- Chlorocalymma

Families of Monocotyledons

- Chrysochloa
- Chrysopogon
- Chumsriella
- Chusquea
- Cinna
- Cladoraphis
- Clausospicula
- Cleistachne
- Cleistochloa
- Cliffordiochloa
- Cockaynea
- Coelachne
- Coelachyropsis
- Coelachyrum
- Coelorachis
- Coix
- Colanthelia
- Coleanthus
- Colpodium
- Commelinidium
- Cornucopiae
- Cortaderia
- Corynephorus
- Cottea
- Craspedorachis
- Cripipes
- Crithopsis
- Crypsis
- Cryptochloa
- Ctenium
- Ctenopsis
- Cutandia
- Cyathopus
- Cyclostachya
- Cymbopogon
- Cymbosetaria
- Cynodon
- Cynosurus
- Cyperochloa
- Cyphochlaena
- Cypholepis
- Cyrtococcum
- Dactylis
- Dactyloctenium
- Daknopholis
- Dallwatsonia
- Danthonia
- Danthoniastrum
- Danthonidium
- Danthoniopsis
- Dasyochloa
- Dasypoa
- Dasypyrum
- Davidsea
- Decaryella
- Decaryochloa
- Dendrocalamus
- Dendrochloa
- Deschampsia
- Desmazeria
- Desmostachya
- Deyeuxia
- Diandrochloa
- Diandrolyna
- Diandrostachya
- Diarrhena
- Dichaetaria
- Dichanthelium
- Dichanthium
- Dichelachne

- Diectomis
- Dielsiochloa
- Digastrium
- Digitaria
- Digitariopsis
- Dignathia
- Diheteropogon
- Dilophotriche
- Dimeria
- Dimorphochloa
- Dinebra
- Dinochloa
- Diplachne
- Diplopogon
- Dissanthelium
- Dissochondrus
- Distichlis
- Drake Brockmania
- Dregeochloa
- Dryopoa
- Dupontia
- Duthiea
- Dybowskia
- Ectoplopus
- Eccoptocarpha
- Echinaria
- Echinochloa
- Echinolaena
- Echinopogon
- Ectrosia
- Ectrosiopsis
- Ehrharta
- Ekmanochloa
- Eleusine
- Eleonurus
- Elymandra
- Elymus
- Elytrigia
- Elytrophorus
- Elytrostachys
- Enneapogon
- Enteropogon
- Entolasia
- Entoplocamia
- Eragrostiella
- Eragrostis
- Eremium
- Eremochloa
- Eremopoa
- Eremopogon
- Eremopyrum
- Eriachne
- Erianthecium
- Erianthus
- Eriochloa
- Eriochrysis
- Erioneuron
- Euchlaena
- Euclasta
- Eulalia
- Eulaleopsis
- Eustachys
- Euthryptochloa
- Exotheca
- Fargesia
- Farrago
- Fasciculochloa
- Festuca
- Festucella
- Festucopsis
- Fingerhuthia
- Froesiochloa
- Garnotia
- Gastridium
- Gaudinia

- → Gaudiniopsis
- → Germainia
- → Gerritea
- → Gigantochloa
- → Glaziochloa
- → Glaziophyton
- → Glyceria
- → Glyphochloa
- → Gouinia
- → Gouldochloa
- → Graphephorum
- → Greslania
- → Griffithsochloa
- → Guaduella
- → Gymnachne
- → Gymnopogon
- → Gynerium
- → Habrochloa
- → Hackelochloa
- → Hainardia
- → Hakonechloa
- → Halopyrum
- → Harpachne
- → Harpochloa
- → Helictotrichon
- → Helleria
- → Hemarthria
- → Hemisorghum
- → Henrardia
- → Hesperostipa
- → Heterachne
- → Heterantheleum
- → Heteranthoecia
- → Heterocarpha
- → Heteropholis
- → Heteropogon
- → Hickelia
- → Hierochloe
- → Hilaria
- → Hitchcockella
- → Holcolemma
- → Holcus
- → Homolepis
- → Homopholis
- → Homozeugos
- → Hookerochloa
- → Hordelymus
- → Hordeum
- → Hubbardia
- → Hubbardochloa
- → Humbertochloa
- → Hyalopoa
- → Hydrochloa
- → Hydrothauma
- → Hygroryza
- → Hylebates
- → Hymenachne
- → Hyparrhenia
- → Hyperthelia
- → Hypogynium
- → Hypseochloa
- → Hystrix
- → Ichnanthus
- → Imperata
- → Indocalamus
- → Indopoa
- → Indosasa
- → Isachne
- → Isalus
- → Ischaemum
- → Ischnochloa
- → Ischnurus
- → Iseilema
- → Ixophorus
- → Jansenella
- → Jardinea
- → Jouvea

- → Joycea
- → Kampochloa
- → Kaokochloa
- → Karroochloa
- → Kengia
- → Kengyilia
- → Kerriochloa
- → Koeleria
- → Lagurus
- → Lamarckia
- → Lamprothyrsus
- → Lasiacis
- → Lasiorhachis
- → Lasiurus
- → Lecomtella
- → Leersia
- → Lepargochloa
- → Leptagrostis
- → Leptaspis
- → Leptocarydion
- → Leptochloa
- → Leptochlopsis
- → Leptocoryphium
- → Leptoloma
- → Leptosaccharum
- → Lepturopetium
- → Leptothrium
- → Lepturella
- → Lepturidium
- → Lepturus
- → Leucophrys
- → Leucopoa
- → Leymus
- → Libyella
- → Limnas
- → Limnodea
- → Limnopoa
- → Lindbergella
- → Linkagrostis
- → Lintonia
- → Lithachne
- → Littledalea
- → Loliolum
- → Lolium
- → Lombardochloa
- → Lophacme
- → Lophatherum
- → Lopholepis
- → Lophopogon
- → Lophopyrum
- → Lorenzochloa
- → Loudetia
- → Loudetiopsis
- → Louisiella
- → Loxodera
- → Luziola
- → Lycochloa
- → Lycurus
- → Lygeum
- → Machurolyra
- → Maillea
- → Malacurus
- → Maltebrunia
- → Manisuris
- → Megalachne
- → Megaloprotachne
- → Megastachya
- → Melanocenchris
- → Melica
- → Melinis
- → Melocalamus
- → Melocanna
- → Merostachys
- → Merxmuellera

Families of Monocotyledons

- → Mesosetum
- → Metasasa
- → Metcalfia
- → Mibora
- → Micraira
- → Microbriza
- → Microcalamus
- → Microchloa
- → Microlaena
- → Micropyropsis
- → Micropyrum
- → Microstegium
- → Mildbraediochloa
- → Milium
- → Miscanthidium
- → Miscanthus
- → Mnesithea
- → Mniochloa
- → Molinia
- → Monachather
- → Monanthochloe
- → Monelytrum
- → Monium
- → Monocladus
- → Monocymbium
- → Monodia
- → Mosdenia
- → Muhlenbergia
- → Munroa
- → Myriocladus
- → Myriostachya
- → Narduroides
- → Nardus
- → Narenga
- → Nassella
- → Nastus
- → Neeragrostis
- → Neesiochloa
- → Nematopoa
- → Neeragrostis
- → Neesiochloa
- → Nematopoa
- → Neobouteloua
- → Neohouzeaua
- → Neostapfia
- → Neostapfiella
- → Nephelochloa
- → Neurachne
- → Neurolepis
- → Neyraudia
- → Notochloe
- → Notodanthonia
- → Ochlandra
- → Ochthochloa
- → Odontelytrum
- → Odyssea
- → Olmeca
- → Olyra
- → Ophiochloa
- → Ophiuros
- → Opizia
- → Oplismenopsis
- → Oplismenus
- → Orcuttia
- → Oreobambos
- → Oreochloa
- → Orinus
- → Oropetium
- → Ortachne
- → Orthoclada
- → Oryza
- → Oryzidium
- → Oryzopsis
- → Otachyrium
- → Otatea
- → Ottochloa
- → Oxychloris

- Oxyrachis
- Oxytenanthera
- Panicum
- Pappophorum
- Parafestuca
- Parahyparrhenia
- Paraneurachne
- Parapholis
- Paratheria
- Parectenium
- Pariana
- Parodiolyra
- Pascopyrum
- Paspalidium
- Paspalum
- Pennisetum
- Pentameris
- Pentapogon
- Pentarrhaphis
- Pentaschistis
- Pereilema
- Periballia
- Peridictyon
- Perotis
- Perrierbambus
- Perulifera
- Petriella
- Peyritschia
- Phacelurus
- Phaenanthoecium
- Phaenosperma
- Phalaris
- Pharus
- Pheidochloa
- Phippsia
- Phleum
- Pholiurus
- Phragmites
- Phyllorhachis
- Phyllostachys
- Pilgerochloa
- Piptatherum
- Piptochaetium
- Piptophyllum
- Pinesia
- Pinesiella
- Plagiantha
- Plagiosetum
- Planichloa
- Plectrachne
- Pleiadelphia
- Pleuropogon
- Plinthanthesis
- Poa
- Pobeguinea
- Podophorus
- Poecilostachys
- Pogonachne
- Pogonanthria
- Pogonatherum
- Pogonachne
- Pogonanthria
- Pogonatherum
- Pogoneura
- Pogonochloa
- Pohlidium
- Poidium
- Polevansia
- Polliniopsis
- Polypogon
- Polytoca
- Polytrias
- Pommereulla
- Porteresia
- Potamophila
- Pringlechloa

- Pseonantheum
- Prosphytochloa
- Psammagrostis
- Psammochloa
- Psathyrostachys
- Pseudanthistiria
- Pseudarrhenatherum
- Pseudechinolaena
- Pseudobromus
- Pseudochaetochloa
- Pseudocoix
- Pseudodanthonia
- Pseudodichanthium
- Pseudopentameris
- Pseudophleum
- Pseudopogonatherum
- Pseudoraphis
- Pseudoroegneria
- Pseudosasa
- Pseudosorghum
- Pseudostachyum
- Pseudovossia
- Pseudoxytenanthera
- Pseudozoysia
- Psilathera
- Psilolemma
- Psilurus
- Pterochloris
- Ptilagrostis
- Puccinellia
- Puelia
- Racemobambos
- Raddia
- Raddiella
- Ratzeburgia
- Radfieldia
- Reederochloa
- Rehia
- Reimarochloa
- Reitzia
- Relchela
- Rendlia
- Reynaudia
- Rhipidocladum
- Rhizocephalus
- Rhomboelytrum
- Rhynchelytrum
- Rhynchoryza
- Rhytachne
- Richardsiella
- Robynsiochloa
- Rottboellia
- Rytidosperma
- Saccharum
- Sacciolepis
- Santedia
- Sasa
- Saugetia
- Schaffnerella
- Schedonnardus
- Schenckochloa
- Schismus
- Schizachne
- Schizachyrium
- Schmidtia
- Schizostachyum
- Schoenefeldia
- Sclerachne
- Sclerochloa
- Sclerodactylon
- Scleropogon
- Sclerostachys
- Scolochloa
- Scribneria
- Scrotochloa
- Scutachne
- Secale

- Sehima
- Semiarundinaria
- Sesleria
- Sesleriella
- Setaria
- Setariopsis
- Shibataea
- Silentvalleya
- Simplicia
- Sinarundinaria
- Sinobambusa
- Sinochasea
- Sitanion
- Snowdenia
- Soderstromia
- Sohnsia
- Sorghastrum
- Sorghum
- Spartina
- Spartochloa
- Spathia
- Sphaerobambos
- Sphaerocaryum
- Spheneria
- Sphenopholis
- Sphenopus
- Spinifex
- Spodiopogon
- Sporobolus
- Steinchisma
- Steirachne
- Stenotaphrum
- Stephanachne
- Stereochlaena
- Steyermarkochloa
- Stiburus
- Stilpnophleum
- Stipa
- Stipagrostis
- Streblochaete
- Streptochaeta
- Streptogyna
- Streptolophus
- Streptostachys
- Styppeiochloa
- Sucrea
- Suddia
- Swallenia
- Swallenochloa
- Symplectrodia
- Taeniatherum
- Taeniorachis
- Tarigidia
- Tatianyx
- Teinostachyum
- Tetrachaete
- Tetrachne
- Tetrapogon
- Tetrarrhena
- Thamnocalamus
- Thaumastochloa
- Thelepogon
- Thellungia
- Themeda
- Thinopyrum
- Thrasya
- Thrasyopsis
- Thuarea
- Thyridachne
- Thyridolepis
- Thyrsia
- Thyrsostachys
- Thysanolaena
- Torreyochloa
- Tovarochloa

- → Tovarochloa
- → Trachypogon
- → Trachys
- → Tragus
- → Triboleum
- → Tricholaena
- → Trichoneura
- → Trichopteryx
- → Tridens
- → Trikeraia
- → Trilobachne
- → Triniochloa
- → Triodea
- → Triplachne
- → Triplasis
- → Triplopogon
- → Tripogon
- → Tripsacum
- → Triraphis
- → Triscenia
- → Trisetum
- → Tristachya
- → Triticum
- → Tsvelevia

- → Tuctoria
- → Uniola
- → Uranthoecium
- → Urelytrum
- → Urochloa
- → Urochondra
- → Vahlodea
- → Vaseyochloa
- → Ventenata
- → Vetiveria
- → Vietnamochloa
- → Vietnamosasa
- → Viguierella
- → Vossia
- → Vulpia
- → Vulpiella
- → Wangenheimia
- → Whiteochloa
- → Willkommia
- → Xerochloa
- → Yakirra
- → Ystia
- → Yushania
- → Yvesia

- → Zea
- → Zenkeria
- → Zeugites
- → Zingeria
- → Zizania
- → Zizaniopsis
- → Zonotreche
- → Zoysia
- → Zygochloa

REFERENCES

References

1. Taxonomy of Angiosperms – AVVS Sambamurthy
2. Modern Plant Taxonomy – N.S. Subramanyam
3. Taxonomy of Angiosperms – B.P. Pandey
4. Taxonomy of Angiosperms – Singh and Jain
5. Introduction to Principles of Plant Taxonomy – Sivaraja
6. Guide to flowering plant families
7. Taxonomy of Angiosperms – Kumaresan
8. Wikipedia
9. www.wildflowers-and-weeds.com
10. www.botany.hawaii.edu
11. www.delta-intkey.com
12. www.theplantlist.org
13. www.wildflowers-and-weeds.com
14. www.britannica.com
15. www.biologydiscussion.com
16. en.m.wikisource.org
17. flora www.eeb.uconn.edu
18. www.babylon-software.com
19. https://eprints.utas.edu.au

Families of Monocotyledons

www.ingramcontent.com/pod-product-compliance
Lightning Source LLC
Chambersburg PA
CBHW071558220526
45469CB00003B/1058